Chemistry
of the
Carbonyl Group

Chemistry
of the
Carbonyl Group

A Programmed Approach

to

Organic Reaction Mechanisms

STUART WARREN

*Department of Organic Chemistry,
Cambridge University, Cambridge*

JOHN WILEY & SONS
Chichester · New York · Brisbane · Toronto · Singapore

Copyright © 1974 John Wiley & Sons Ltd. All Rights Reserved. No part of the publication may be reproduced, stored in a retrieval system, or transmitted, in any form or by any means, electronic, mechanical, photocopying, recording or otherwise, without the prior written permission of the Copyright owner.

Library of Congress Catalog Card No. 74-6701
ISBN 0 471 92104 1

Reprinted January 1981, March 1983,
February 1985, October 1987,
August 1990, September 1992,
August 1994, December 1997,
September 2000

Printed and bound in Great Britain by
Antony Rowe Ltd, Chippenham and Eastbourne

PREFACE

This program was originally written for a lecture course given to second year university students. I thought that the course needed some backing in the form of down-to-earth material for the students to use themselves, so that they had to translate new ideas into their own thoughts, expressed in their own way and reinforced by the experience of using the ideas to solve problems. The students were enthusiastic about the program and so I have now rewritten it for more general use hoping that it will be helpful to anyone taking a university or college course, studying from a book, revising for an exam or simply brushing up their organic chemistry.

The program is an *aid* to learning: a piece of support material which should help the student in his work but does not stand by itself. Lecturers, "advisers", and teachers will probably find the pages describing the contents and concepts for each section useful in assessing which part of a course the program best supports.

I should like to thank Andrew Crompton for working through the program and making many perceptive suggestions, Hilary Rhodes for typing the printed words and Howard Jones for his encouragement and practical help in producing the book.

Cambridge 1974   Stuart Warren

CONTENTS

Some help you may need

What do you need to know before you start?

How to use the Program

The Program

Contents of Section 1

Concepts Assumed and Concepts Introduced in Section 1

Section 1 : Nucleophilic Addition to the Carbonyl Group (frames 1-65)

Contents of Section 2

Concepts Assumed, Introduced, and Reinforced in Section 2

Section 2: Nucleophilic Substitution at the Carbonyl Group (frames 66-120)

Contents of Section 3

Concepts Assumed, Introduced, and Reinforced in Section 3

Section 3: Nucleophilic Substitution at the Carbonyl Group with Complete Removal of Carbonyl Oxygen (frames 121-168)

Contents of Section 4

Concepts Assumed, Introduced, and Reinforced in Section 4

Section 4: Carbanions and Enolisation (frames 169-216)

Concepts Assumed, Introduced, and Reinforced in Section 5

Section 5: Building Organic Molecules from Carbonyl Compounds (frames 217-330)

SOME HELP YOU MAY NEED

This program is to help you learn about the chemistry of the carbonyl group. I am assuming that you are using the program as part of a more comprehensive piece of learning such as a University course. You should therefore have access to a library or at least one reliable textbook of organic chemistry and an "adviser"; that is someone such as a tutor or fellow student with whom you can discuss chemistry. You may well need both these as you work through the program. To help you fill in the background I'll give references to chapters in some of the more helpful texts:

"Cram"
J.B. Hendrickson, D.J. Cram, and G. Hammond; Organic Chemistry, 3rd ed., McGraw Hill 1970.

"Norman"
R.O.C. Norman; Principles of Organic Synthesis, Methuen, 1968.

"Roberts"
J.D. Roberts and M. Caserio; Basic Principles of Organic Chemistry, Benjamin, 1964.

"Sykes"
P. Sykes; A Guidebook to Mechanism in Organic Chemistry, 2nd ed., Longmans, 1965.

"Tedder"
J.M. Tedder and A. Nechvatal; Basic Organic Chemistry, Wiley (Vols 1-4) 1966.

WHAT DO YOU NEED TO KNOW BEFORE YOU START?

The program is about only a small part of organic chemistry and I have to assume that you know certain facts and appreciate certain concepts. In case you feel uncertain about any of these, here is a list of them with appropriate remedies in each case.

I assume you can:

- Draw and recognise structures of simple organic compounds (an aldehyde, acetone, n-butanol...). Any organic text will tell you about this.

- Write mechanisms using curly arrows (e.g. to described the S_N2 reaction between hydroxide ion and methyl iodide). If you can't do this you should read Cram ch 9, Norman ch 4, Roberts ch 7 sec 4, Sykes p 13-19, or Tedder vol 1 ch 3, or consult your adviser.

- Outline the basic chemistry of alkyl halides including nucleophilic attack at saturated carbon (substitution and elimination reactions S_N1, S_N2, E1 and E2 mechanisms). A brush up with Cram chs 10 and 14, Norman ch 4, Roberts ch 11, Sykes chs 3 and 8, or Tedder vol 1 chs 3, 4, and 5 might help.

- Explain what is meant by: the periodic table, pK_a value, anion and cation, electrophile and nucleophile, lone pair electrons; tetrahedral structure, σ- and π-bonds, intramolecular and reversible reactions. Again, any text will help.

WHAT DO YOU NEED TO KNOW BEFORE YOU START (contd.)

This list applies to the whole program but at the beginning of each section you will find a list of "concepts assumed". This will be a more detailed analysis of the concepts I expect you to have grasped before you start that section. You will find help in the textbooks or from your adviser if you are unsure of any of these concepts.

There are also lists of "concepts introduced" and "concepts reinforced" for each section. You may find these useful as a check list to make sure, after you have finished the section, that you really have met and understood these concepts. In many cases a concept introduced in one section will be assumed for the next.

HOW TO USE THE PROGRAM

Remember at all times that the point of program learning is that you <u>learn at your own pace</u> and that you yourself <u>check on your own progress</u>. I shall give you information and ideas in chunks called frames, each numbered and separated by a black line. Normally each frame includes a question which is sometimes followed by a comment or a clue, and always by the answer. A program is an active form of learning: if it is to be of any good to you, you must play your part by actually <u>writing down the answers to the questions</u> as you go along and checking up on points you aren't sure about.

When you are ready to start, cover the first page with a card and pull it down to expose the first frame. Read and act on that frame and then expose frame 2, and so on. Remember to write down the answers to the questions - they are for you to check on your own progress and it is often only when you commit yourself to paper that you find out whether you really understand what you are doing.

I hope you find the program enjoyable and helpful.

Nucleophilic Addition, 1:1

CONTENTS OF SECTION I: NUCLEOPHILIC ADDITION
                         TO THE CARBONYL GROUP

Frame 2 Nucleophilic addition: what it is and how it happens.
  8 Alcohols as nucleophiles: acetal formation.
 22 Some carbon-carbon bond forming reactions with carbon nucleophiles: cyanide ion, acetylide ion and Grignard reagents.
 41 Hydride ion and its derivatives $LiAlH_4$ and $NaBH_4$. Reduction of aldehydes and ketones.
 47 Meerwein-Pondorff reduction and Oppenauer oxidation, with a branch program on how to draw transition states.
 62 Two general revision problems.

Nucleophilic Addition, 1:2

Concepts Assumed:

σ- and π-bonds. Inductive effects.
Polarisable bonds. pK_a values.
Electrophile and Periodic table.
 nucleophile. Transition states.
Conjugation with
 π-bonds and lone pairs.

Concepts Introduced:

Acid catalysis.
Instability of $R_2C(OR)_2$ in acid solution.
Driving equilibria in a chosen direction by the use of acid, solvent etc.
Stability of different carbonyl compounds.
Stability and reactivity as two sides of the same coin.
Effect of substituents on equilibria.
Relationship between basicity and nucleophilicity.
Organo-magnesium compounds as nucleophiles.
Use of Grignard reagents in syntheses.
Ease of dehydration of t-alcohols in acid solution.
Sources of nucleophilic H^-.
Use of Al and B compounds an anion transferring reagents.
Drawing transition states.
Stability of the six-membered ring.
Use of protecting groups (name not used).
Use of reaction mechanisms in syntheses.

Nucleophilic Addition, 1:3

SECTION I: NUCLEOPHILIC ADDITION TO THE CARBONYL GROUP

Have you read the introductions explaining what help you need, what you need to know, and how to use the program? It's a good idea to do this before you start.

---

1. The carbonyl group (2a) has an easily polarisable π-bond with an electrophilic carbon atom at one end easily attacked by nucleophiles:

$$\overset{\backslash +}{\underset{/}{C}} = \overset{-}{O} \quad (2a) \qquad \underset{X^-}{\overset{\backslash}{\underset{/}{C}} = \overset{\curvearrowright}{O}} \quad \rightleftharpoons \quad \overset{\backslash}{\underset{/}{C}} \overset{O^-}{\underset{X}{\diagdown}}$$

Write down the reaction (with curly arrows) between acetone and hydroxide ion.

---

2. Have you actually written down the formulae of the reagents and drawn the arrows? The program won't be of much help to you unless you do.

---

3. $\underset{CH_3}{\overset{CH_3}{\diagdown}} \underset{\underset{HO^-}{\overset{\curvearrowleft}{}}}{C {\overset{\curvearrowright}{=}} O} \quad \rightleftharpoons \quad \underset{CH_3}{\overset{CH_3}{\diagdown}} C \overset{O^-}{\underset{OH}{\diagdown}}$

A nucleophile such as water uses its lone pair electrons (··) to attack and forms a neutral addition compound by proton transfer:

$$\underset{\overset{\cdot\cdot}{O}H_2}{\overset{\backslash}{\underset{/}{C}} \overset{\curvearrowright}{=} O} \rightleftharpoons \overset{\backslash}{\underset{/}{C}} \overset{O^-}{\underset{+OH_2}{\diagdown}} \rightleftharpoons C \overset{OH}{\underset{OH}{\diagdown}}$$

Note that only one proton is needed. Write down the reaction between the carbonyl compound acetaldehyde and the proton-bearing

Nucleophilic Addition, 1:4

nucleophile ethanol.

4. If you find this difficult, use the lone pair electrons on the ethanol oxygen atom to attack the carbonyl group of acetone.

5. Answer to frame 3:

$$CH_3-\underset{Et\ddot{O}H}{\overset{H}{\underset{|}{C}}}=\ddot{O} \rightleftharpoons CH_3-\underset{\underset{EtH}{\overset{+}{O}}}{\overset{H}{\underset{|}{C}}}-\bar{O} \rightleftharpoons CH_3-\underset{EtO}{\overset{H}{\underset{|}{C}}}-OH$$

Another approach is to add the proton first, in acid solution, and to add the nucleophile afterwards:

$$\underset{}{\overset{}{>}}C=\ddot{O}\overset{H^+}{\rightleftharpoons} \underset{}{\overset{+}{>}}C=\overset{+}{O}H \underset{XH}{\rightleftharpoons} \underset{\overset{+}{X}H}{>}C\overset{OH}{\underset{}{<}} \rightleftharpoons \underset{X}{>}C\overset{OH}{\underset{}{<}} + H^+$$

In this case the proton is regenerated and this is an example of acid catalysis. Show how water can be added to acetone with acid catalysis.

6. If you are having difficulty with this, look back at the last reaction in frame 5. Carry out these same steps using acetone as the carbonyl compound and water as HX.

7.

$$\underset{CH_3}{\overset{CH_3}{>}}C=\ddot{O}\overset{H^+}{\rightleftharpoons} \underset{CH_3}{\overset{CH_3}{>}}\underset{\overset{\cdot\cdot}{O}H_2}{\overset{+}{C}-OH} \rightleftharpoons \underset{CH_3}{\overset{CH_3}{>}}C\overset{OH}{\underset{\overset{+}{O}H_2}{<}} \rightleftharpoons \underset{CH_3}{\overset{CH_3}{>}}C\overset{OH}{\underset{OH}{<}}$$

Notice that $Me_2C=OH^+$ is much more reactive than acetone, but is still attacked at <u>carbon</u> although the positive charge is in fact on

Nucleophilic Addition, 1:5

the oxygen atom. What would happen to
$R_2C=OH^+$ with water, ethanol, $PhCH_2SH$, and
cyanide ion?

8.

$R_2C\begin{matrix}OH\\OH\end{matrix}$ $R_2C\begin{matrix}OH\\OEt\end{matrix}$ $R_2C\begin{matrix}OH\\SCH_2Ph\end{matrix}$ $R_2C\begin{matrix}OH\\CN\end{matrix}$

8b

would be formed by a mechanism exactly like
the one in frame 7.
When you combined $R_2C=OH^+$ with ethanol you
formed an adduct (8b) which is the product of
ethanol addition to a ketone. The steps you
drew are therefore part of the acid catalysed
addition of ethanol to a ketone. Draw out
the whole of this reaction.

9.

$R_2C=O \underset{}{\overset{H^+}{\rightleftharpoons}} R_2C\text{-}\overset{+}{O}H \rightleftharpoons R_2C\begin{matrix}OH\\OEt\end{matrix} \rightleftharpoons R_2C\begin{matrix}OH\\OEt\end{matrix}$
$\qquad\qquad\quad Et\overset{..}{O}H \qquad\qquad H^+$

---

10. Look at the reactions in frames 7 and 9
again. Note that all the steps are reversible
and that therefore $R_2C=OH^+$ may be formed from
$R_2C=O$, $R_2C(OH)_2$, or $R_2C(OH)OR$:

$R_2C=O \xrightarrow{H^+} R_2C=OH^+ \longleftarrow R_2C\begin{matrix}OH\\ \overset{+}{O}R\\H\end{matrix}$

10c

It is in fact a general rule that compounds of
the type $R_2C(OR)_2$, having two oxygen atoms
singly bonded to the same carbon atom, are
unstable in acid solution. A reason for this
is that both oxygens have lone pairs of

Nucleophilic Addition, 1:6

electrons, and so when one pair is protonated, the lone pair on the other can form a C=O double bond and expel the protonated atom. Draw this in detail.

11.

$$R_2C\begin{smallmatrix}\ddot{O}R\\OR\end{smallmatrix} \xrightarrow{H^+} R_2C\begin{smallmatrix}\overset{+}{O}R\\H\\OR\end{smallmatrix} \rightleftharpoons R_2C=\overset{+}{O}R + ROH$$

Look at the reactions in frame 10 again. In the reverse reaction we protonated and removed the EtO- group from 10c. What happens if we protonate and remove the HO- group?

12.

$$R_2C\begin{smallmatrix}\ddot{O}H\\OR\end{smallmatrix} \xrightarrow{H^+} R_2C\begin{smallmatrix}\overset{+}{O}H_2\\OR\end{smallmatrix} \rightleftharpoons R_2C=\overset{+}{O}R$$

This new cation, $R_2C=OEt^+$ is just as reactive as $R_2C=OH^+$ and can add nucleophiles in the same way. What happens if we add EtOH to it?

13.

$$R_2\overset{+}{C}=\underset{Et\ddot{O}H}{OEt} \rightleftharpoons R_2C\begin{smallmatrix}OEt\\\overset{+}{O}Et\\H\end{smallmatrix} \rightleftharpoons R_2C\begin{smallmatrix}OEt\\OEt\end{smallmatrix}$$

This reaction sequence, added to the ones in frames 9 and 12 gives us the addition of two molecules of ethanol to a ketone to give $R_2C(OEt)_2$. Draw this sequence out in full without referring back.

---

14. If you have difficulty doing this, look at frames 9, 12, and 13 without writing anything down and then try.

Nucleophilic Addition, 1:7

15.

$$R_2C=O \underset{}{\overset{H^+}{\rightleftarrows}} R_2\overset{+}{C}-OH \atop Et\ddot{O}H \rightleftarrows R_2C\overset{OH}{\underset{\overset{+}{O}Et}{\diagdown}} \rightleftarrows R_2C\overset{OH}{\underset{OEt}{\diagdown}}$$
$$H$$

$$R_2C\overset{OH}{\underset{OEt}{\diagdown}} \overset{H^+}{\rightleftarrows} R_2C\overset{\overset{+}{O}H_2}{\underset{OEt}{\diagdown}} \rightleftarrows R_2\overset{+}{C}-\overset{..}{O}Et \atop Et\ddot{O}H \rightleftarrows R_2C\overset{OEt}{\underset{\overset{+}{O}Et}{\diagdown}}$$
$$H$$

$$\rightleftarrows R_2C\overset{OEt}{\underset{OEt}{\diagdown}}$$

---

16. Does your reaction sequence exactly follow that in frame 15? If not, try to assess if the differences are trivial. If you are still in doubt, consult your adviser. It is important that you understand this reaction well. This is a good place to stop if you want a break.

---

17. Since this whole sequence is reversible, it will go forwards in ethanol and backwards in water. What do you think would happen if acetaldehyde and n-butanol were dissolved together in a 1:3 molar ratio, and the solution refluxed for twelve hours with a catalytic quantity of toluene sulphonic acid, and the product dried and distilled?

---

18. This is a literature preparation of $CH_3CH(OBu-n)_2$. Acetaldehyde is the carbonyl component, butanol the nucleophile, and toluene sulphonic acid is the catalyst. How would you hydrolyse $PhCH(OEt)_2$, and what would you get?

Nucleophilic Addition, 1:8

19. Reflux the PhCH(OEt)$_2$ in water with a catalytic quantity of an acid. The products would be PhCHO and EtOH. How would you make EtCH(OMe)$_2$?

20. Treat EtCHO and MeOH as in frame 17. A glance at the reactions in frame 15 should convince you that the <u>mono</u> adducts of carbonyl compounds and nucleophiles with lone pairs are unstable. An example is R$_2$C(OH)OEt which is unstable even under the conditions of its formation. What happens to it?

21. In ethanol it gives the acetal R$_2$C(OEt)$_2$, in water the carbonyl compound R$_2$C=O.

22. If we look instead at nucleophiles <u>without</u> lone pairs we should find some stable mono adducts. Which of the adducts in frame 8 should be stable?

23. The cyanide substituent has no lone pair and so its adduct, R$_2$C(OH)(CN) should be stable. These compounds, cyanohydrins, are made by adding excess NaCN and one mole of acid to the carbonyl compound. The reaction is an equilibrium. What is the role of the acid?

24. To drive over the equilibrium by protonating the intermediate:

$$R_2C=O \;+\; ^-CN \;\rightleftharpoons\; R_2C(O^-)(CN) \;\xrightarrow{H^+}\; R_2C(OH)(CN)$$

Nucleophilic Addition, 1:9

25. Since this reaction is an equilibrium, the amount of cyanohydrin formed from any given carbonyl compound will depend on the relative stabilities of the carbonyl compound itself and the product. There can be many substituents X on a carbonyl compound R.CO.X, such as Cl, Me, NH_2, Ph, OEt, H. Some have inductive effects, some conjugate with the carbonyl group. Some stabilise RCOX making it less reactive. Others activate it towards nucleophilic attack. Arrange the compounds RCOX, where X can be the substituents listed above, into an order of reactivity towards a nucleophile.

---

26. Before you look at the answer in the next frame, just check that you have considered each of these factors: Some substituents stabilise RCOX by π-conjugation:

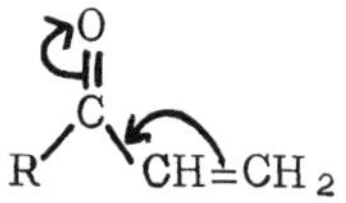

Some by lone pair conjugation:

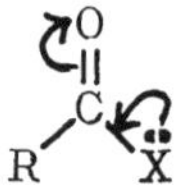

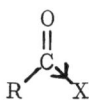

Some <u>destabilise</u> RCOX by inductive electron withdrawal:

O
‖
R—C—X

Nucleophilic Additon, 1:10

27. Taking RCHO as standard, we can say that Cl <u>de</u>stabilises by inductive withdrawal, Me stabilises more by π-delocalisation, and NH_2 and OEt by lone pair donation (NH_2 is more effective at this: compare ammonia and water as bases). Our order is:

Cl...H...Me...Ph...OEt...NH_2
most reactive   most stable

These same factors could affect the product as well. Arrange the same substituents in an order for product stability.

28. Since the carbonyl group has gone in $R_2C(OH)CN$, we would expect very little effect from any of these substitutents. There can't be any conjugation, and inductive effects on the distant O atom will be small. What effect would an inductively withdrawing substituent have on the equilibrium for cyanohydrin formation?

29. It will destabilise the carbonyl compound a lot, and have very little effect on the product: it will therefore push the equilibrium over to the cyanohydrin side. Consider cyanohydrin formation from:

a) MeCHO   c) $Me_2C=O$ e) $Ph_2C=O$
b) PhCHO   d) PhCO.Me   f) MeCOOEt

Two of these give no cyanohydrin at all. Which two? One gives 100 per cent cyanohydrin, which? When equilibrium is established, c forms 10,000 times as much cyanohydrin as d. Comment.

Nucleophilic Addition, 1:11

30. -e and f give no cyanohydrin at all.
-a gives all cyanohydrin.
-in d there is a strong π-conjugation from the benzene ring absent in c which stabilises the carbonyl compound. These results fit in with what we have said, if you don't see, consult your adviser.

---

31. Another carbon anion you may have met is the acetylide ion, formed by the action of strong base on acetylene:

$$H-C\equiv C-H \quad \curvearrowleft {}^-NH_2 \longrightarrow H-C\equiv C^-$$

It adds readily to the carbonyl group. Draw out the reaction.

32.
$$R_2C=O \atop {}^{\curvearrowleft}C\equiv C-H \longrightarrow R_2\underset{C\equiv CH}{\overset{|}{C}}-\bar{O} \xrightarrow{H^+} R_2\underset{C\equiv CH}{\overset{|}{C}}-OH$$

---

33. The pK_a of HCN is 9.15, that of water is 15.7, and that of acetylene about 25. Which anion, CN^-, HO^-, or acetylide ion, would add fastest to acetone?

---

34. Acetylide fastest, then hydroxide, then cyanide slowest. Remember that if HX is a weak acid, X^- is a strong base.

---

35. Another type of carbon nucleophile is the Grignard reagent RMgBr made by direct metalation of the organic halide with magnesium metal. These compounds are nucleophilic through carbon because the electrons in the C-Mg bond polarise towards carbon. Draw the attack of MeMgBr on $CH_2=O$.

Nucleophilic Addition, 1:12

36.

$$BrMg\frown CH_3 \searrow CH_2\!=\!\overset{\frown}{O} \longrightarrow CH_3-CH_2-\overset{-}{O} \;\; \overset{+}{MgBr}$$
$$\longrightarrow CH_3CH_2O-MgBr$$

The intermediate O-Mg compound hydrolyses in acid solution by attack of water on the magnesium atom. Suggest how this might occur.

---

37. A possible mechanism is:

$$EtO-MgBr \rightarrow EtO-\overset{+}{\underset{..}{Mg}} \overset{\frown}{\underset{..}{O}H_2} \rightarrow EtO-Mg-\overset{+}{\underset{..}{O}H_2} \rightleftharpoons$$

$$\underset{+}{Et\overset{H}{\underset{|}{O}}—MgOH} \rightarrow EtOH$$

In any event, the product is a primary alcohol and is a general route from RBr to RCH$_2$OH. What would be the reaction between PhCH$_2$MgBr and CH$_3$CHO?

38.

$$BrMg\overset{Ph}{\underset{\frown}{\underset{|}{CH_2}}} \overset{H}{\underset{|}{\underset{CH_3}{C\!=\!\overset{\frown}{O}}}} \longrightarrow \overset{Ph}{\underset{|}{\underset{CH_3}{\underset{|}{CH_2}}}}\!\!\!\!\!\!CH\!-\!\overset{-}{O} \;\; MgBr \longrightarrow Ph\!\!\bigwedge\!\!\!\!\!\!\!\!\!\overset{OMgBr}{}$$

$$\xrightarrow[H_2O]{H^+} Ph\!\!\bigwedge\!\!\!\!\!\!\!\!\!\overset{OH}{}$$

So we can make secondary alcohols this way. Ketones also react with Grignard reagents. Draw the reaction between PhCH$_2$MgBr and Ph$_2$C=O.

39.

$$BrMg\overset{Ph}{\underset{\frown}{\underset{|}{CH_2}}} \overset{Ph}{\underset{|}{\underset{Ph}{C\!=\!\overset{\frown}{O}}}} \longrightarrow \overset{Ph}{\underset{|}{CH}}\!\!\!\overset{Ph}{\underset{Ph}{\underset{|}{\overset{|}{C}-O^-}}} \xrightarrow[H_2O]{H^+} Ph\!\!\bigwedge\!\!\!\!\!\!\!\!\overset{Ph\; Ph}{\underset{OH}{}}$$

The work-up is done in acid solution, and tertiary alcohols react easily with acid. What further reaction might happen here?

Nucleophilic Addition, 1:13

40. Protonation and loss of water leads to a stable tertiary carbonium ion:

Ph-CH₂-C(Ph)(Ph)-OH + H⁺ → Ph-CH₂-C(Ph)(Ph)-OH₂⁺ → Ph₂C=CHPh + Ph₂C⁺-CH₃ + H·

So much for carbon nucleophiles. This is a good place to rest.

---

41. The simplest nucleophile of all is the hydride ion, H⁻, but as you may know, this ion is very basic and will not add to the carbonyl group. Sources of H⁻ for addition to the C=O bond are NaBH₄ and LiAlH₄ containing the tetrahedral anions BH₄⁻ and AlH₄⁻. Draw out the structure of these ions.

42.

H-B⁻(H)(H)H   H-Al⁻(H)(H)H

Show how AlH₄⁻ may transfer H⁻ as a nucleophile to a ketone R₂C=O.

43.

H-Al⁻(H)(H)H + R₂C=O → H-C(R)(R)-O⁻ →[H⁺] R₂CH-OH

LiAlH₄, lithium aluminium hydride is dangerous when damp. Can you suggest why?

---

44. It gives off hydrogen in large volumes (H⁻ + H₂O → H₂ + OH⁻) and evolves heat at the same time. The result is usually an impressive fire. Sodium borohydride, NaBH₄

Nucleophilic Addition, 1:14

is less reactive and can be used in alkaline aqueous solution.

What would be formed from PhCOMe and NaBH₄?

45.

$$H_3B^- + Ph-\underset{CH_3}{\underset{|}{C}}=O \longrightarrow H-\underset{CH_3}{\underset{|}{C}}-O^- \xrightarrow{H_2O} \underset{CH_3}{\overset{Ph}{\diagdown}}CH-OH$$

These reagents demonstrate two important properties of boron and aluminium compounds. If you are uncertain of the periodic table just check to see where these two elements come. Neutral tervalent B and Al compound are electron deficient:

$$\underset{R}{\overset{R}{\diagdown}}B-R \qquad \text{only six valency electrons: no lone pair.}$$

They readily accept nucleophiles to form stable tetravalent anions. Draw the reaction between BF₃ and a fluoride ion, F⁻.

46. $F^- \curvearrowright BF_3 \longrightarrow F\!-\!\!\bar{\ }BF_3$

Also, anions can be transferred from these tetrahedral anions to other molecules:

$$R_3B\!\frown\!X \quad \downarrow \quad Y^+ \longrightarrow R_3B + X-Y$$

These two properties may be summarised by saying that tervalent boron and aluminium compounds will accept anions from one molecule and transfer them to another.

47. This is used in the selective reduction

Nucleophilic Addition, 1:15

technique known as the Meerwein-Pondorff reduction. The tervalent compound is aluminium iso-propoxide, $Al(OPr-i)_3$. When a compound such as a ketone is added to this reagent, it combines with it to form a tetrahedral anion. Draw this.

---

48. If you are in difficulty, remember that the Al atom is electrophilic and therefore combines with nucleophiles, and think which end of the ketone molecule is <u>nucleo</u>philic.

49.
$$(i-PrO)_3Al \quad \ddot{O}=C\overset{R}{\underset{R}{\diagdown}} \longrightarrow (i-PrO)\bar{A}l-\overset{+}{O}=C\overset{R}{\underset{R}{\diagdown}}$$

If we draw this same intermediate with one of the iso propyl groups drawn out in full we can see that the tertiary hydrogen atom in the iso-propyl group (*) is electronically and geometrically placed so that it can be transferred to the ketone. Put arrows on the formula to show how this happens.

$$(i-PrO)_2\bar{A}l \diagup^{\overset{+}{O}=CR_2}_{O-\underset{CH_3}{\underset{|}{C}}\diagdown^{H^*}_{CH_3}} \longrightarrow (i-PrO)_2\bar{A}l \diagup^{O-CR_2}_{\underset{H}{|}} \quad + \quad O=C(CH_3)_2$$

50.
$$(i-PrO)_2\bar{A}l \diagup^{\overset{+}{O}=CR_2}_{O-\underset{CH_3}{\underset{|}{C}}\diagdown^{H}_{CH_3}} \longrightarrow (i-PrO)_2Al-OCHR_2 \quad + \quad Me_2C=O$$

---

51. The products so far are acetone and a

Nucleophilic Addition, 1:16

new aluminium compound. The reaction is done in isopropanol solution so that another i-PrO group displaces the newly formed ketone from the Al atom:

$(i\text{-PrO})_2\text{Al OCHR}_2 \underset{}{\overset{i\text{-PrOH}}{\rightleftarrows}} (i\text{-PrO})_3\text{Al} + \text{R}_2\text{CHOH}$

The reaction is done at a high enough temperature for the acetone to distil off and the equilibrium is kept over to the right.

---

52. In the hydrogen transfer reaction (frame 50), how is the hydrogen actually transferred, as an atom, a proton, or a hydride ion?

---

53. As a hydride ion, that is with the pair of electrons from the C-H bond. This type of reaction is called a hydride transfer. You may have noticed that the transfer was done intra-molecularly within a six-membered ring and therefore has a six-membered transition state. Draw it.

---

54. If you do know how to draw transition states, skip to frame 59. If you don't know, read on. The transition state for a reaction is the state of highest energy along the reaction path. To draw it, one must first know the reaction mechanism in detail. If there is more than one step, then there will be a transition state for each step. We shall go

Nucleophilic Addition, 1:17

through the process for a one step reaction, BH_4^- and acetone. Draw the mechanism for this reaction.

55.

$$H_3B^-\frown H \quad \overset{CH_3}{\underset{CH_3}{>}}C\overset{\frown}{=}O \longrightarrow H_3B \;+\; H-\underset{CH_3}{\overset{CH_3}{|}}C-O^-$$

Now draw the formulae again, but put in only those bonds which remain unaffected by the reaction. Don't draw charges or arrows.

56.

$$H_3B \qquad H \qquad \overset{CH_3}{\underset{CH_3}{>}}C-O$$

Now "dot" in all bonds which are formed or broken during the reaction, and mark all appearing or disappearing charges in brackets to show that they are partial charges. (Don't use δ+ or δ- as these mean very small charges and here the total charge must add up to unity).

57.

$$H_3\overset{(-)}{B}\text{-----}H\text{---}\overset{CH_3}{\underset{CH_3}{>}}C=O^{(-)}$$

Now draw the transition state for this reaction:

$$\underset{R}{\overset{O}{\underset{\diagdown}{C}}}\overset{\|}{\underset{O}{\diagup}} \overset{\curvearrowleft}{H} \longleftarrow CH_2\overset{+}{=}\overset{-}{N}=N \longrightarrow \underset{R}{\overset{\bar{O}}{\underset{\diagdown}{C}}}\overset{|}{\underset{O}{\diagup}} \;+\; CH_3-\overset{+}{N}\equiv N$$

Nucleophilic Addition, 1:18

58.

$$\underset{R}{\overset{(-)}{\underset{O}{\parallel}}{C}} \underset{O}{\overset{}{\diagdown}} H\text{---}CH_2 \text{===} \overset{+}{N} \text{===} N^{(-)}$$

If this wasn't clear, consult your adviser. If it was clear, you ought to be able to do the original problem so go back to frame 53.

59.

$$(i\text{-PrO})_2 Al \overset{(-)}{\diagdown} \begin{matrix} O\text{===}CR_2 \\ \vdots \\ H \\ \vdots \\ O\text{===}C \\ | \quad \diagdown CH_3 \\ CH_3 \end{matrix} \overset{(+)}{}$$

The reason that this reaction goes so well is that six-membered rings are very stable, and so an intramolecular reaction going through a six-membered transition state will be most favourable.

---

60. The reverse reaction is known as the Oppenauer oxidation. Here aluminium tri tertiary butoxide and an involatile ketone such as cyclohexanone are used to oxidise any secondary alcohol to the corresponding ketone.

$$(t\text{-BuO})_3 Al \xrightarrow{R_2CHOH} (t\text{-BuO})_2 AlOCHR_2$$

$$\longrightarrow (t\text{-BuO})_2^- Al \diagdown \begin{matrix} O\text{---}\overset{R}{\underset{H}{C}}\text{---}R \\ \\ O\text{===}\bigcirc \\ + \end{matrix}$$

Show how a hydride transfer in this intermediate leads to the oxidation of the alcohol.

Nucleophilic Addition, 1:19

61.

$(t\text{-BuO})_2\text{Al}^- \text{—O—C}{\overset{R}{\underset{H}{\diagdown}}}\text{R} \quad \longrightarrow \quad (t\text{-BuO})_2\text{Al—O—}\underset{\text{(cyclohexyl)}}{\text{C}_6\text{H}_{11}} + R_2C=O$

(with cyclohexanone coordinated at Al)

So we can reduce a ketone to an alcohol or oxidise an alcohol to a ketone by using $\text{Al}(\text{OPr-i})_3$ and acetone in the one case and $\text{Al}(\text{OBu-t})_3$ and cyclohexanone in the other. By their mechanism you can see that these reactions will have no effect on other functional groups such as C=C double bonds.

---

62. This is nearly the end of the first part of the program, so here are some general problems.

You will remember that acetal formation (frames 7-17) is a reversible reaction. It turns out that the equilibrium constant for acetal formation from a ketone is unfavourable:

$$R_2CO + R'OH \underset{}{\overset{H^+}{\rightleftharpoons}} R_2C(OR')_2$$

and poor yields are obtained. However cyclic acetals can be made from ethylene glycol. Draw out the mechanism for this reaction:

$$(CH_3)_2C=O + HOCH_2CH_2OH \xrightarrow{H^+} \underset{CH_3}{\overset{CH_3}{\diagup}}C\underset{O-CH_2}{\overset{O-CH_2}{\diagdown}}\Big|$$

---

63. If you need some help, begin by adding one HO group of $HOCH_2CH_2OH$ to the protonated ketone, just as you did with ethanol in frame 5.

Nucleophilic Addition, 1:20

64.

[Reaction scheme showing acid-catalysed acetal formation of acetone with ethylene glycol, proceeding through protonated acetone, hemiketal, protonated hemiketal, ring closure to protonated cyclic hemiketal, and finally the cyclic acetal (1,3-dioxolane of acetone).]

Predict what happens in this reaction sequence:

$$CH_3\overset{O}{\underset{\|}{C}}CH_2\underset{\underset{CH_3}{|}}{\overset{\overset{CH_3}{|}}{C}}CHO \xrightarrow{EtOH} A \xrightarrow[iPrOH]{Al(OPr\text{-}i)_3} B \xrightarrow[H_2O]{H^+} C$$

---

65. In the first step the acetal from the aldehyde group but not from the ketone group (frame 62) is formed:

$$CH_3CO.CH_2CMe_2CHO \longrightarrow CH_3CO.CH_2CMe_2CH(OEt)_2$$

Only one carbonyl group is now available for Meerwein-Pondorff reduction (frames 47-61).

$$CH_3CO.CH_2CMe_2CH(OEt)_2 \longrightarrow$$
$$CH_3CH(OH).CH_2CMe_2CH(OEt)_2$$

Finally the acetal is hydrolysed by standard means. The result of all this is that we have reduced a ketone in the presence of an aldehyde:

$$\underset{CH_3}{}\overset{O}{\underset{\|}{C}}\diagdown\overset{Me\diagdown\diagup Me}{\underset{CH_2}{}\underset{}{C}}\diagdown CHO \longrightarrow \underset{CH_3}{}\overset{OH}{\underset{}{CH}}\diagdown\overset{Me\diagdown\diagup Me}{\underset{CH_2}{}\underset{}{C}}\diagdown CHO$$

This is the end of the first part of the program.

NUCLEOPHILIC SUBSTITUTION, 2.1

CONTENTS OF SECTION II: NUCLEOPHILIC SUB-
STITUTION AT THE
CARBONYL GROUP

Frame 66 Substitution: how it happens.
      76 LiAlH$_4$ reduction of esters.
      80 Reaction of Grignard reagents with esters.
      89 Alkaline hydrolysis of esters.
      94 Acid hydrolysis of amides.
      98 Summary of acid and base catalysis.
     102 Reaction between carboxylic acids and thionyl chloride.
     110 Synthesis of esters and anhydrides from carboxylic acids.
     115 Review questions.

NUCLEOPHILIC SUBSTITUTION, 2.2

SECTION II: NUCLEOPHILIC SUBSTITUTION AT THE CARBONYL GROUP

Concepts assumed:

Electrostatic repulsion.

Catalysis.

Heavy isotopes.

Acidity and basicity of Cl^-, EtO^-, NH_3, etc.

Concepts introduced:

Relationship between nucleophilicity, leaving group ability, and basicity.

Substitution as an extension of addition.

Position and use of electrophilic attack on carboxylic acid derivatives.

Similarity of reaction between cyclic and acyclic compounds.

Why carboxylic acids are unreactive towards nucleophilic substitution.

Use of ^{18}O in establishing reaction mechanisms.

Concepts reinforced:

Use of acid and base catalysis and choice of solvent in driving an equilibrium in the chosen direction.

Advantages of the six-membered cyclic transition state.

Use of Grignard reagents in synthesis.

Ease of dehydration of tertiary alcohols.

Use of reaction mechanisms in designing syntheses.

NUCLEOPHILIC SUBSTITUTION, 2.3

66. In the first part of the program we considered only aldehydes and ketones. We're now going to look at the full range of structures including carboxylic acids, RCO.OH, acid chlorides RCO.Cl, and esters RCO.OR. Using your knowledge of nucleophilic addition, what would be the first reaction between hydroxide ion and benzoyl chloride PhCO.Cl?

67.

$$\begin{array}{c} O \\ \parallel \\ Ph-C-Cl \\ HO^- \end{array} \longrightarrow \begin{array}{c} O^- \\ | \\ Ph-C-Cl \\ | \\ OH \end{array}$$

Instead of picking up a proton from the solvent, this intermediate has a better reaction: the negative charge returns to restore the C=O double bond expelling Cl⁻. Draw this.

68.

$$\begin{array}{c} O^- \\ | \\ Ph-C-Cl \\ | \\ OH \end{array} \longrightarrow \begin{array}{c} O \\ \parallel \\ Ph-C-OH \end{array} + Cl^-$$

Why did it expel Cl⁻ and not OH⁻?

---

69. Because Cl⁻ is a better leaving group than OH⁻. We know this because HCl is a stronger acid than HOH and is therefore readier to ionise. Cl⁻ must therefore be more stable than OH⁻.

Draw out the whole of the reaction.

NUCLEOPHILIC SUBSTITUTION, 2.4

70.

$$\text{Ph-C(=O)-Cl} + \text{HO}^- \rightleftharpoons \text{Ph-C(O}^-\text{)(OH)-Cl} \rightarrow \text{Ph-C(=O)-OH} + \text{Cl}^-$$

This is then a substitution reaction, OH in the product taking the place of Cl in the reactant.

---

71. What happens if aniline, $PhNH_2$, and benzoyl chloride react together?

---

72. Aniline is the nucleophile:

$$\text{Ph-C(=O)-Cl} + \text{PhNH}_2 \rightleftharpoons \text{Ph-C(O}^-\text{)(Cl)-NH}_2^+\text{Ph} \rightleftharpoons \text{Ph-C(O}^-\text{)(Cl)-NHPh} \rightarrow \text{Ph-C(=O)-NHPh} + \text{Cl}^-$$

Cl^- is preferred to $PhNH^-$ as leaving group. Which is the stronger nucleophile towards a carbonyl group, Cl^- or OH^-?

---

73. HO^- because it is more basic.
It is in fact a general principle that as far as the carbonyl group is concerned, molecules are better nucleophiles if they are more _____ and worse leaving groups if they are more _____.

---

74. **basic** fills both gaps.
To put it another way, a more/less basic nucleophile will displace a more/less basic leaving group from a carbonyl compound. (Choose 'more' or 'less' in each case.)

---

75. more. less.
Which of these reactions would you expect to work well? (See next page)

NUCLEOPHILIC SUBSTITUTION, 2.5

a) $Cl^- + CH_3CONH_2 \longrightarrow CH_3CO.Cl + NH_2^-$
b) $NH_3 + CH_3CO.OEt \longrightarrow CH_3CO.NH_2 + EtOH$
c) $CH_3CO.O^- + CH_3CO.Cl \longrightarrow CH_3CO.OCO.CH_3 + Cl^-$

---

76. b) and c) will work, a) won't. If you don't see this, consult your adviser. What do you think will happen here?

$$RCO.OEt + LiAlH_4 \longrightarrow$$

---

77. If you have problems, read frames 42 and 43 to remind yourself that the AlH_4^- ion delivers H^- as a nucleophile to the carbonyl group.

78.

$$H_3\overline{Al}{-}H \curvearrowright \overset{R}{\underset{OEt}{C}}{=}O \longrightarrow H{-}\overset{R}{\underset{OEt}{C}}{-}O^- \longrightarrow R{-}\overset{O}{\underset{\|}{C}}{-}H + EtO^-$$

EtO^- is less basic than H^- and so is displaced. What will happen to the product under the reaction conditions?

---

79. The aldehyde is reduced to the alcohol:

$$H_3\overline{Al}{-}H \curvearrowright \overset{R}{\underset{H}{C}}{=}O \longrightarrow H{-}\overset{R}{\underset{H}{C}}{-}O^- \longrightarrow RCH_2OH$$

Since the aldehyde is normally more reactive than the ester, it is virtually impossible to stop at the first stage and the compound is normally treated with an excess of $LiAlH_4$ to give the alcohol.

NUCLEOPHILIC SUBSTITUTION, 2.6

80. A very important series of nucleophiles are the Grignard reagents RMgBr. What would happen with RCO.OEt and R'MgBr? The reaction begins like this:

$$BrMg-R' \rightarrow \underset{OEt}{\overset{R}{C}}=O$$

---

81. Have you considered the possibility of further reaction? Frames 35 - 38 may help.

---

82. As before, the first reaction converts these carbonyl compounds into reactive products, this time ketones. An excess of Grignard reagent therefore gives good yields of tertiary alcohols.

$$BrMg-R' \rightarrow \underset{OEt}{\overset{R}{C}}=O \rightarrow R'-\overset{R}{C}=O \leftarrow R'-MgBr \rightarrow R'-\underset{OH}{\overset{R}{\underset{|}{C}}}-R'$$

The double headed arrow on the ester molecule is a useful shorthand to indicate the two stages of the substitution reaction. We shall use it from now on, and if you are in doubt about its meaning you should consult your adviser.

The final product from this reaction was a tertiary alcohol. How might this be made into an olefin?

---

83. Tertiary alcohols give carbonium ions with great ease in acid solution (see frames 39 - 40) and the carbonium ion can either

NUCLEOPHILIC SUBSTITUTION, 2.7

pick up a nucleophile or lose a proton to give an olefin. To get a good yield of olefin you therefore want an acid with a weakly nucleophilic anion such as H_2SO_4 or $KHSO_4$. In fact this reaction often happens during the work-up of the Grignard reaction.

$$CH_3-\underset{OH}{\underset{|}{\overset{CH_2CH_3}{\overset{|}{C}}}}-CH_3 \xrightarrow{H^+} CH_3-\underset{\overset{+}{O}H_2}{\underset{|}{\overset{CH_2CH_3}{\overset{|}{C}}}}-CH_3 \rightarrow \underset{CH_3 \quad CH_3}{\overset{H\quad CHCH_3}{\overset{|}{\overset{+}{C}}}}$$

$$\rightarrow \underset{CH_3 \quad CH_3}{\overset{H \quad CH_3}{C=C}}$$

---

84. What product would be formed in this reaction:

cyclohexyl-CO.OEt $\xrightarrow[\text{2. } H^+, H_2O]{\text{1. PhMgBr}}$ A $\xrightarrow{KHSO_4}$ B

---

85. Here we used a cyclic substituent.

cyclohexyl-CO.OEt $\xrightarrow{PhMgBr}$ cyclohexyl-CO-Ph $\xrightarrow{PhMgBr}$ cyclohexyl-C(OH)(Ph)(Ph) A $\xrightarrow{KHSO_4}$ cyclohexylidene=CPh$_2$ B

Now try this reaction which has a cyclic ester or lactone in it. The reaction is just like that for an ordinary ester.

isochroman-1-one $\xrightarrow[\text{2. } H^+, H_2O]{\text{1. PhCH}_2\text{MgBr}}$ A $\xrightarrow{KHSO_4}$ B

NUCLEOPHILIC SUBSTITUTION, 2.8

86.

[isochroman-1-one] → [2-(2-hydroxyethyl)phenyl benzyl ketone, with OH and CH₂Ph] →

[diol: Ph—C(OH)(CH₂Ph)—Ar—CH₂CH₂OH] → [alkene: Ph—C(=CHCH₃)—Ar—CH₂CH₂OH, with Ph]

---

87. A Grignard reaction is probably an example of electrophilic catalysis involving two molecules of the Grignard reagent:

$$R_2C{=}O \quad MgR' \quad \longrightarrow \quad R_2C\text{—}OMgBr$$
$$R' \quad Mg \quad Br \qquad\qquad R' \quad Mg\text{—}Br$$
$$Br \qquad\qquad\qquad\qquad\qquad Br$$

This intramolecular mechanism may remind you of the Meerwein-Pondorff reduction (frames 49 - 59). Both have a six-membered cyclic transition state. Draw it for this reaction.

88.

$$R_2C{\equiv}O$$
$$R' \quad\quad Mg\text{—}R'$$
$$Mg\text{---}Br$$
$$Br$$

This is a good place to rest if you want to.

---

89. In view of our conclusion in frame 74, it may surprise you to recall that esters are hydrolysed in alkaline solution:

$$CH_3\text{—}\underset{O}{\overset{O}{\|}}{C}\text{—}OEt + HO^- \xrightarrow{H_2O} CH_3\text{—}\underset{O}{\overset{O}{\|}}{C}\text{—}OH + EtOH$$

Draw a mechanism for this.

NUCLEOPHILIC SUBSTITUTION, 2.9

90.

$$HO^- \curvearrowright \underset{OEt}{\overset{CH_3}{C}}=O \rightleftharpoons HO-\underset{OEt}{\overset{CH_3}{\underset{|}{C}}}-O^- \rightleftharpoons HO-\overset{CH_3}{\underset{|}{C}}=O$$

$$+ \; EtO^- \rightleftharpoons EtOH$$

Here OH^- displaces the slightly more basic EtO^-. Can you think why this happens so easily?

---

91. A clue: did you notice that the steps are all equilibria? Think how an equilibrium can be driven over, and perhaps compare this reaction with the hydrolysis of acetals (frame 17).

---

92. The reaction is done in water with an excess of OH^- to drive the reaction forward by the mass action effect. There is another reason too. Think which species will actually be formed in aqueous alkali.

---

93. CH_3COOH will form CH_3COO^- and electrostatic repulsion will prevent attack of OH^- on this so that it is effectively removed from the equilibrating system.

---

94. Amides can also be hydrolysed in alkali, but let's look at their hydrolysis in acid solution:

$$CH_3CO.NH_2 + H_2O \xrightarrow[H^+]{} CH_3COOH + NH_3$$

How could acid catalyse the first step of the reaction, the formation of the tetrahedral

NUCLEOPHILIC SUBSTITUTION, 2.10

intermediate?

$$R-\underset{\parallel}{\overset{O}{C}}-NH_2 \xrightarrow{H_2O} R-\underset{OH}{\overset{OH}{\underset{|}{C}}}-NH_2$$

---

95. By protonation. Note that we use the lone pair of electrons on nitrogen, but protonate at the carbonyl group to get the most delocalised cation:

$$R-\overset{H^+}{\underset{NH_2}{\diagup}} \longrightarrow R-\overset{OH}{\underset{H_2O:}{\underset{NH_2}{\diagup}}}{}^+ \longrightarrow R-\underset{+OH_2}{\overset{OH}{|}}-NH_2$$

$$\longrightarrow R-\underset{OH}{\overset{OH}{|}}-NH_2$$

None of these substituents are now good leaving groups, but in acid solution one of them might be protonated. Which?

---

96. NH_2 is most basic:

$$R-\underset{OH}{\overset{OH}{|}}-NH_2 \quad \underset{}{\overset{H^+}{\rightleftarrows}} \quad R-\underset{OH}{\overset{OH}{|}}-\overset{+}{N}H_3$$

Now we have a good leaving group. Draw the next step.

---

97. $$R-\underset{:OH}{\overset{OH}{|}}-\overset{+}{N}H_3 \longrightarrow R-\underset{\overset{+}{OH}}{\overset{OH}{\diagdown}} \rightleftarrows R-\underset{}{\overset{O}{\parallel}}-OH + NH_3$$

This reaction is very similar to some of the steps in acetal formation (frames 10-12).

---

98. So to summarise, base can catalyse the

NUCLEOPHILIC SUBSTITUTION, 2.11

the hydrolysis of esters or amides in two ways. State them.

---

99. 1: By driving over an unfavourable equilibrium using the mass action effect.
2: By capturing the carboxylic acid product as an unreactive anion.

---

100. Acid catalyses the same reaction also in two ways. State them.

---

101. 1: It catalyses the addition of the nucleophile by protonating the carbonyl group.
2: It turns what is otherwise a bad leaving group into a good one by protonation.

---

102. This first catalytic function (101:1) can be carried out by electrophiles other than the proton. This applies particularly to carboxylic acids. The electrophile could attack either oxygen atom:

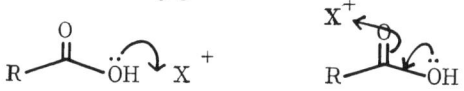

Which atom do you think will actually be attacked?

---

103. The carbonyl oxygen atom will be attacked because the cation produced is delocalised over both oxygen atoms:

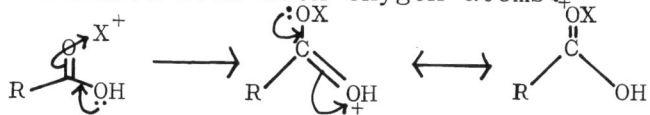

A good example of this kind of reaction is the attack of thionyl chloride on carboxylic

NUCLEOPHILIC SUBSTITUTION, 2.12

acids. Initial attack occurs at carbonyl oxygen. Draw the products.

R–C(Ö:OH)(–O–S(O)(O)–Cl) ⟶ [product with S–Cl, C–Cl]

104.

[structure: R–C(=OH⁺)–O–S(=O)–Cl]   Cl⁻

We now have within the system:
    a) a reactive carbonyl group. Why?
    b) a good leaving group. Which?
    c) a nucleophile. What?

105. a) Because it is protonated.
b) $ClSO_2^-$ which in fact decomposes to SO_2 and Cl^-.
c) Chloride ion.

So what do you think happens now?

106.

[mechanism: protonated mixed anhydride with Cl⁻ attacking] ⟶ R–C(=O)–Cl + ⁻O–S(=O)–Cl ⟶ SO_2 + Cl⁻

The net result is that a carboxylic acid has been converted into an acid chloride.

107. You may have noticed that this is the first example of a nucleophilic substitution at a carboxylic acid that we have seen, and we find in general that, unless we attack first with an electrophile, carboxylic acids are very unreactive towards nucleophilic substitution. Can you think of a reason for this?

NUCLEOPHILIC SUBSTITUTION, 2.13

108. These are really related reasons:
The carbonyl group is rather unreactive.
The leaving group would have to be HO^-, notoriously one of the worst.
Perhaps the most important of the three is that nucleophiles are bases and they therefore remove the acidic proton rather than attack the carbonyl group.
How do we overcome these problems by reaction with thionyl chloride (frames 103 - 106)?

---

109. The carbonyl group has become the reactive $C=OH^+$. We have a very good leaving group in $ClSO_2^-$. The removal of a proton has become an irrelevance since it's got to come off anyway and because of the electrophilic assistance we can use a very weak nucleophile, Cl^-.
Other reagents which do the same job are $POCl_3$ and PCl_5.

---

110. How might we convert RCO.OH to RCO.OEt?

---

111. We know that EtOH will displace Cl^- from the carbonyl group with base catalysis, so we need to make RCO.Cl, and this we have just done. The whole scheme is:

RCO.OH $\xrightarrow{SOCl_2}$ RCO.Cl $\xrightarrow[\text{base}]{\text{EtOH}}$ RCO.OEt

In fact we don't always need to go through RCO.Cl, as acid catalysed reaction between the acid and EtOH often gives the ester in good yield. Mechanism?

NUCLEOPHILIC SUBSTITUTION, 2.14

112. Protonation gives the most delocalised cation (frame 95):

$$R-C(OH)_2^+ \rightleftharpoons R-C(OH)(O^+HEt) \rightleftharpoons R-C(OH)(OEt) \rightleftharpoons R-C(OEt)(O^+H_2)$$

$$\rightleftharpoons R-C(^+OH)(OEt) \rightleftharpoons RCO.OEt + H^+$$

This is another example of an equilibrium, so we make an ethyl ester from RCO.OH, EtOH and acid in solution in _____, and we hydrolyse an ester in _____ with acid catalysis.

---

113. We make the ester in <u>ethanol</u>, and hydrolyse it in <u>water</u>. Another useful compound is the anhydride RCO.OCO.R. How could we make that from RCO.OH using reactions we've discussed?

---

114. Again we need to displace Cl^- by $RCO.O^-$:

$$RCO.OH \xrightarrow{SOCl_2} RCO.Cl$$
$$RCO.OH \xrightarrow{NaHCO_3} RCO.O^-$$
$$\longrightarrow RCO.O.COR$$

This is practically the end of the second section, so here are some general review questions.

---

115. Arrange these compounds in order of reactivity towards water; all reactions are to give the carboxylic acid:

$CH_3CO.OCO.CH_3$ $CH_3CO.NH_2$
$CH_3CO.Cl$ $CH_3CO.OEt$

NUCLEOPHILIC SUBSTITUTION, 2.15

116. Chloride ⟩ anhydride ⟩ ester ⟩ amide. In fact the chloride explodes with cold water, the anhydride reacts with cold water, the ester reacts with dilute acid or base but the amide only with boiling 70% acid or 10% caustic soda. Ask you adviser if you are in doubt about this.

---

117. If you hydrolysed an ester labelled with heavy oxygen:

$$CH_3-\overset{O}{\underset{\|}{C}}-{}^{18}O-Et$$

in acid or base, would the ^{18}O end up in the acetic acid or the ethanol?

---

118. In both acid and base it would end up in the ethanol, this is one of the pieces of evidence used to establish the mechanisms we have been discussing. If you want to read more about this see Cram p.504, Tedder vol 3, ch.6.

$$CH_3-\overset{O^-}{\underset{HO}{\overset{|}{C}}}-{}^{18}O-Et \rightleftharpoons CH_3-\overset{O}{\underset{OH}{\overset{\|}{C}}}-{}^{18}O-Et \rightleftharpoons CH_3CO.OH + H^{18}OEt$$

---

119. How would you carry out this multi-step synthesis?

$$PhCH_2CO.OH \longrightarrow PhCH_2CH_2O.COCH_2Ph$$

---

120. $PhCH_2CO_2H \xrightarrow[H^+]{EtOH} PhCH_2CO_2Et \xrightarrow{LiAlH_4} PhCH_2CH_2OH$

$PhCH_2CO_2H \xrightarrow{SOCl_2} PhCH_2CO.Cl \longrightarrow PhCH_2CH_2OCO.CH_2Ph$

There are other good routes: if you have one discuss it with your adviser. This is the end of the second part of the program.

REMOVAL OF CARBONYL OXYGEN, 3:1

CONTENTS OF SECTION III: NUCLEOPHILIC SUB-
                          STITUTION AT THE
                          CARBONYL GROUP WITH
                          COMPLETE REMOVAL OF
                          CARBONYL OXYGEN

Frame   121   Imine formation from aldehydes
              and ketones.
        127   Oxime formation and the structure
              of oximes.
        130   Hydrazone and semicarbazone form-
              ation.
        137   Reduction of C=O to CH_2.
        149   Conversion of C=O to CCl_2.
        154   DDT synthesis.
        158   Chloromethylation of aromatic
              compounds.
        165   Review questions.

REMOVAL OF CARBONYL OXYGEN, 3.2

SECTION III: NUCLEOPHILIC SUBSTITUTION AT THE CARBONYL GROUP WITH COMPLETE REMOVAL OF CARBONYL OXYGEN

Concepts Assumed

Rigidity of olefin two dimensional structure.
Geometrical isomerism.
Mechanism of the E2 reaction.

Concepts Introduced

Absence of acid chloride-like substitution in aldehydes and ketones.
Possibility of loss of carbonyl oxygen from tetrahedral intermediate.
Geometrical isomerism of oximes.
Usefulness of different reagents using contrasting conditions for the same synthetic step.
Non-nucleophilicity of amide nitrogen atoms.
Use of high-boiling solvents.
Medium ring compounds.
Characterising a compound as a stable crystalline derivative.

Concepts Reinforced

Instability of compounds containing two atoms, both with lone pairs, bonded to the same carbon atom.
Incompatibility of strongly basic nucleophile and strongly acidic conditions.
Electrophilic attack of S and P compounds on carbonyl oxygen.
Electrophilic substitution in the benzene ring.

REMOVAL OF CARBONYL OXYGEN, 3.3

SECTION III: NUCLEOPHILIC SUBSTITUTION AT THE CARBONYL GROUP LEADING TO COMPLETE REMOVAL OF THE OXYGEN ATOM.

121. You may have the impression from the last section that aldehydes and ketones can't do substitution reactions, but this isn't true. We don't of course get reactions like this:

$$CH_3-\underset{\underset{}{\overset{\overset{O}{\|}}{C}}}{}-CH_3 + HO^- \longrightarrow CH_3-\underset{\underset{}{\overset{\overset{O}{\|}}{C}}}{}-OH + CH_4$$

Why not?

---

122. The mechanism would have to be:

$$CH_3-\underset{HO^-}{\overset{\overset{O}{\|}}{C}}\overset{\curvearrowleft}{-}CH_3 \rightarrow CH_3-\underset{OH}{\overset{\overset{O^-}{|}}{C}}-CH_3 \rightarrow CH_3CO_2H + CH_3^- \xrightarrow{H_2O} CH_4$$

and CH_3^- is far too basic to be displaced. We are going to look at a new kind of substitution reaction. Draw out the addition of aniline, $PhNH_2$, to acetone to give a neutral adduct.

123.

$$CH_3-\underset{Ph\overset{..}{N}H_2}{\overset{\overset{O}{\|}}{C}}\overset{\curvearrowleft}{-}CH_3 \longrightarrow CH_3-\underset{\underset{Ph}{\overset{}{\overset{+}{N}H_2}}}{\overset{\overset{O^-}{|}}{C}}-CH_3 \longrightarrow CH_3-\underset{\underset{Ph}{\overset{}{NH}}}{\overset{\overset{OH}{|}}{C}}-CH_3$$

If this addition reaction is to be extended into a substitution, we must find a leaving group. Neither CH_3, OH, nor PhNH are good leaving groups but either OH or PhNH can become so. How?

REMOVAL OF CARBONYL OXYGEN, 3.4

124. By protonation (as in amide hydrolysis, frames 94 - 97). If PhNH is protonated and eliminated, we just reverse the reaction back to the starting materials, but see what happens if you protonate and eliminate OH. Draw the reaction.

125.

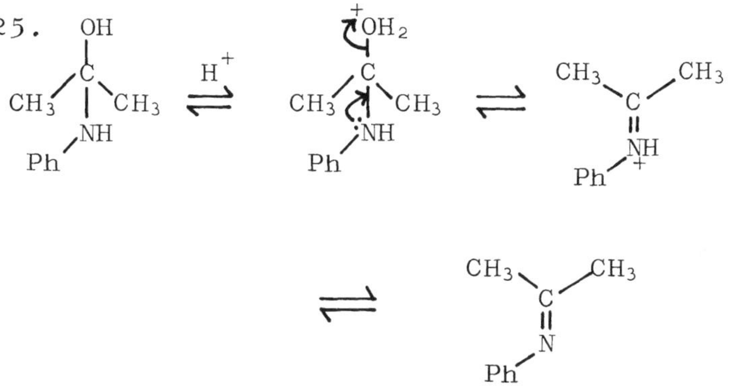

CH₃–C(OH)(CH₃)–NHPh ⇌ (H⁺) CH₃–C(⁺OH₂)(CH₃)–NHPh ⇌ CH₃–C(=NH⁺Ph)–CH₃ ⇌ CH₃–C(=NPh)–CH₃

What you have done is to make the carbonyl oxygen atom the leaving group: the reactions we are going to explore in this section all involve complete removal of carbonyl oxygen during a substitution reaction.

---

126. The product of this last reaction is an imine, containing a C=N double bond. The formation of these compounds is an equilibrium and they are very easily hydrolysed. More stable imines are formed from hydroxyl-amine HONH₂. Write down the complete reaction between this compound and PhCO.Me (using the N atom as nucleophile).

REMOVAL OF CARBONYL OXYGEN, 3.5

127.

$$Ph-\underset{O:}{\overset{\|}{C}}-CH_3 \;\; \underset{\ddot{N}H_2OH}{\rightleftarrows} \;\; Ph\underset{\underset{OH}{\overset{+}{N}H_2}}{\overset{CH_3}{\diagup}}O^- \;\; \rightleftarrows \;\; Ph\underset{\underset{OH}{NH}}{\overset{CH_3}{\diagup}}OH$$

$$\rightleftarrows \;\; Ph\underset{\underset{OH}{\ddot{N}H}}{\overset{CH_3}{\diagup}}\overset{+}{O}H_2 \;\; \rightleftarrows \;\; Ph\underset{\underset{H}{\overset{+}{N}OH}}{\overset{CH_3}{\diagup}} \;\; \rightleftarrows \;\; Ph\underset{NOH}{\overset{CH_3}{\diagup}}$$

Draw out the full (two dimensional) structure of this product.

128.
$$\underset{CH_3}{\overset{Ph}{\diagup}}C=N\diagdown^{OH}$$

How many isomers of this compound are there?

---

129. There are in fact two: the C=N bond is just as rigid as the C=C double bond, and these compounds, oximes, are like olefins with one substituent missing: there are "cis" and "trans" forms:

$$\underset{CH_3}{\overset{Ph}{\diagup}}{=}\ddot{N}\diagdown^{OH} \qquad \underset{CH_3}{\overset{Ph}{\diagup}}{=}\ddot{N}\diagdown_{OH}$$

---

130. Similar reactions occur with other amines, particularly hydrazine, NH_2NH_2, and its derivatives which form good stable crystalline imines known as hydrazones. What compound would be formed from hydrazine and acetone?

131.
$$\underset{CH_3}{\overset{CH_3}{\diagup}}{=}O \;\; + \;\; NH_2NH_2 \;\; \overset{H^+}{\longrightarrow} \;\; \underset{CH_3}{\overset{CH_3}{\diagup}}{=}N-NH_2$$

Supposing there were more acetone around, what might happen now?

REMOVAL OF CARBONYL OXYGEN, 3.6

132. The other end of the molecule could react too:

$$\text{(CH}_3)_2\text{C=N-NH}_2 + \text{O=C(CH}_3)_2 \xrightarrow{H^+} (\text{CH}_3)_2\text{C=N-N=C(CH}_3)_2$$

---

133. Semicarbazide, $NH_2NHCONH_2$, also reacts well: what product would be formed here with acetaldehyde?

134.

$$\underset{H}{\overset{CH_3}{>}}C=O + NH_2.NHCONH_2 \xrightarrow{H^+} \underset{H}{\overset{CH_3}{>}}C=N.NHCONH_2$$

This nitrogen atom is in fact the only one to react. Why don't the other two react as well?

---

135. The other two nitrogen atoms have their lone pairs conjugated to the carbonyl group: they are in fact amide-like. It is only the terminal nitrogen atom which is fully nucleophilic.

$$\ddot{N}H_2\text{---}\ddot{N}H\text{---}\overset{\overset{O}{\|}}{C}\text{---}\ddot{N}H_2$$

---

136. We have met the hydride ion in the guise of $NaBH_4$ and $LiAlH_4$ twice already (frames 41-45 and 76 - 79). If we now use hydride ion to remove the carbonyl oxygen atom altogether we shall obviously get a hydrocarbon:

$$R_2C=O \longrightarrow R_2CH_2$$

What could be an intermediate in this reaction

REMOVAL OF CARBONYL OXYGEN, 3.7

137. The obvious intermediate is found by adding hydride ion to the carbonyl compound in the usual way:

$$H^- \curvearrowright R_2C=O \longrightarrow R_2C(H)(O^-) \longrightarrow R_2C(H)(OH)$$

Under strongly acidic conditions this alcohol could give a carbonium ion. Draw this.

138.

$$R_2C(OH)(H) \xrightarrow{H^+} R_2C(OH_2^+)(H) \rightleftharpoons R_2CH^+$$

All we have to do now is to add hydride ion to the carbonium ion to get the hydrocarbon. You may have been nursing a growing feeling that all is not well with this idea, and you are right. What is wrong?

---

139. If you're not sure, consider whether there isn't something incompatible with the nature of the hydride ion, or $NaBH_4$ or $LiAlH_4$, and the conditions we have outlined for this reaction.

---

140. We need strong acid to protonate and eliminate the carbonyl oxygen atom, and we can't possibly use any of the sources of hydride ion in acid solution: $NaBH_4$ would react violently to give hydrogen gas, $LiAlH_4$ or NaH would explode.

We therefore use a dissolving metal reduction in strong acid. This reaction, the Clemmensen reduction, may use the principle we have outlined here, but its mechanism is unknown in detail. $R_2C=O + Zn/Hg + $ conc. $HCl \longrightarrow R_2CH_2$

REMOVAL OF CARBONYL OXYGEN, 3.8

141. An alternative method begins with the formation of a hydrazone. We shall use cyclohexanone here for a change. Draw the product formed from this ketone and hydrazine (the hydrazone).

142.

cyclohexanone + NH$_2$NH$_2$ / H$^+$ → cyclohexanone hydrazone (=N-NH$_2$)

The hydrazone has all the elements of the product we want, cyclohexane, plus two nitrogen atoms. All(!) we have to do is to move two hydrogen atoms from nitrogen to carbon. This can be done in very strong base:

[mechanism: hydrazone + OH⁻ ⇌ intermediate → =N=N-H species]

Complete the reaction mechanism to give cyclohexane.

143.

[mechanism steps ending with cyclohexane + N$_2$]

The reaction is usually done at high temperatures in ethylene glycol, a high-boiling polar solvent, and is called the Wolff-Kishner reduction:

$$R_2C=O + NH_2NH_2 + NaOH \xrightarrow[200°]{\text{reflux in glycol}} R_2CH_2 + N_2$$

---

144. Yet another method is to make the dithioacetal from the ketone, say PhCO.CH$_2$CH$_3$, and HSCH$_2$CH$_2$SH. Draw this.

REMOVAL OF CARBONYL OXYGEN, 3.9

145. If you can't do this, look back at the frames on cyclic acetal formation (62-3): it's really the same reaction.

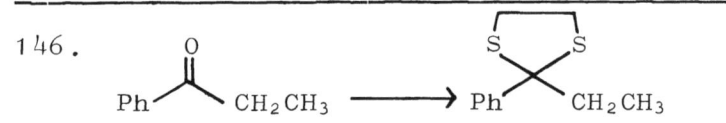

---

146. Ph-CO-CH₂CH₃ ⟶ Ph-C(S-S)-CH₂CH₃ (dithioketal, 5-membered S-C-S ring)

The dithioketal can be reduced directly to the hydrocarbon by hydrogenation over the sulphur-removing catalyst Raney nickel:

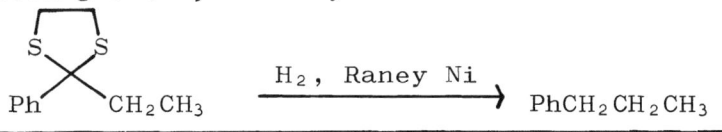

Ph-C(S-S)-CH₂CH₃ —H₂, Raney Ni→ PhCH₂CH₂CH₃

---

147. We now have methods of reducing ketones to hydrocarbons in acidic, alkaline, and neutral solutions. This is useful since we may have molecules which are sensitive to some of these conditions. Which methods would you use for these reactions?

a) PhCO.CMe₂.CH(OEt)₂ ⟶ PhCH₂CMe₂CH(OEt)₂

b) ortho-disubstituted benzene with COCH₃ and CONH₂ ⟶ ortho-disubstituted benzene with CH₂CH₃ and CONH₂

---

148. a) Wolff-Kishner. Clemmensen would destroy the acetal.
b) Dithioketal. Either of the other methods would do nasty things to the amide. This is a good place to stop if you'd like a rest.

---

149. The idea of adding a nucleophile to a carbonyl compound to give an alcohol and then dehydrating the alcohol to a carbonium ion in acid solution, and finally adding another

REMOVAL OF CARBONYL OXYGEN, 3.10

nucleophile to the carbonium ion, could be extended a bit. In frames 137-8 we discussed this idea for H$^-$ as nucleophile, but it didn't work out too well. Draw the sequence out for Cl$^-$ as nucleophile on a general ketone R$_2$C=O.

150.

$$R_2C=O \;\; Cl^- \longrightarrow R_2C\begin{matrix}OH\\Cl\end{matrix} \xrightarrow{H^+} R_2C\begin{matrix}\overset{+}{O}H_2\\Cl\end{matrix} \longrightarrow R_2\overset{+}{C}-Cl \;\; Cl^-$$

$$\longrightarrow R_2C\begin{matrix}Cl\\Cl\end{matrix}$$

It turns out that we can't do this with HCl, we need a stronger electrophilic catalyst for the first step than the proton. We use instead PCl$_5$, which reacts by ionising to Cl$^-$ and PCl$_4{}^+$. The products are R$_2$CCl$_2$, and POCl$_3$. See if you can complete the mechanism.

---

151. $R_2C=\ddot{O} \;\; {}^+PCl_4 \longrightarrow R_2C=\overset{+}{O}-PCl_4 \longrightarrow R_2\overset{|}{C}-O-PCl_4$
$\phantom{151. \;\; R_2C=\ddot{O} \;\; {}^+PCl_4 \longrightarrow R_2C=\overset{+}{O}-PCl_4 \longrightarrow R_2}Cl$

$R_2\overset{+}{C}-Cl \;\; + \;\; {}^-O-PCl_3-Cl \longrightarrow O=PCl_3 + Cl^-$

$Cl^- \longrightarrow R_2CCl_2$

Draw the product you would get from PCl$_5$ on PhCO.CH$_2$Ph.

---

152. Ph.CCl$_2$.CH$_2$Ph.
Supposing you now reacted this with EtO$^-$ in EtOH. What would you get?

---

153. Stuck? Ethoxide ion is very basic and likes to remove protons, even protons attached to carbon atoms. We also have good leaving groups in the molecule and that combination

REMOVAL OF CARBONYL OXYGEN, 3.11

looks like making an elimination reaction...

154.

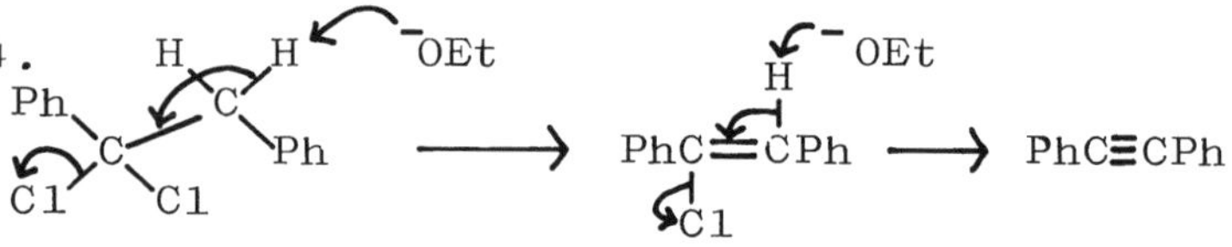

Ph₂C(Cl)–CH(Ph)H + ⁻OEt → PhC=CPh (with H, Cl leaving) → PhC≡CPh

A double elimination by the E2 mechanism.

---

155. At this point you have a choice. I want to deal with some substitutions on the benzene ring using mechanisms like that for the nitration of benzene. If you're quite familiar, read on. If you want to brush up on the subject, turn to Cram ch. 16. If you know nothing about the subject, either skip to frame 165 or read Cram ch. 16 and do your best.

---

156. With very reactive carbonyl compounds, such as chloral Cl₃C.CHO, we can even add aromatic compounds in strong acid. Draw a mechanism for the first step, the formation of the alcohol:

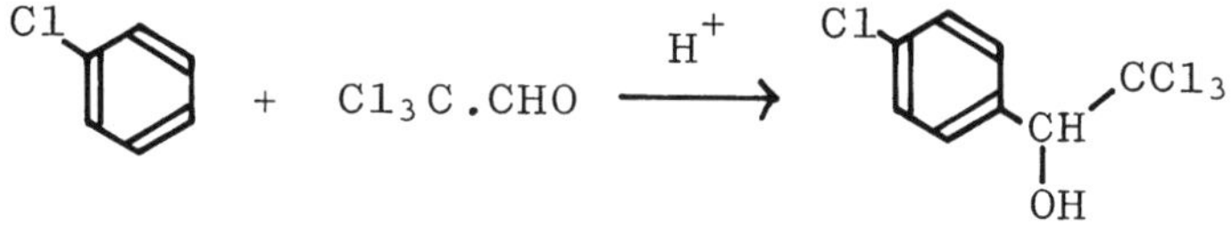

PhCl + Cl₃C.CHO —H⁺→ 4-Cl-C₆H₄-CH(OH)-CCl₃

157.

Cl₃C.CHO —H⁺→ Cl₃C-CH=⁺OH → Cl₃C-CH(OH)-C₆H₄Cl (+H) → Cl₃C.CH(OH)-C₆H₄Cl

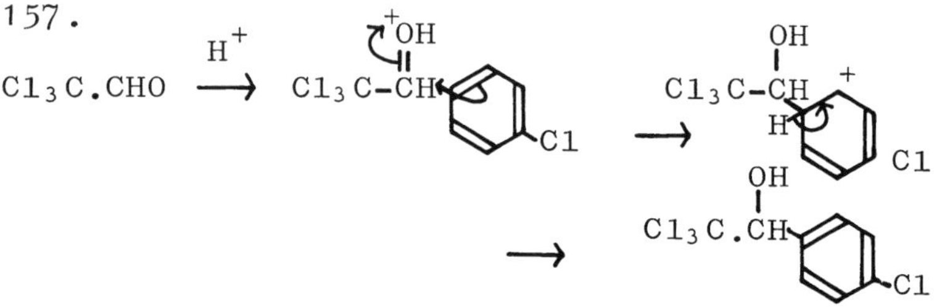

Now a second molecule adds on:

REMOVAL OF CARBONYL OXYGEN, 3.12

Cl–C₆H₄–CH(OH)–CCl₃ + C₆H₅Cl →(H⁺) Cl–C₆H₄–CH(CCl₃)–C₆H₄–Cl

Suggest a mechanism for this step.

---

158. Cl–C₆H₄–CH(⁺OH₂)–CCl₃ → Cl–C₆H₄–⁺CH–CCl₃ + C₆H₅Cl → Cl–C₆H₄–CH(CCl₃)–C₆H₄(H)(⁺)–Cl

Cl–C₆H₄–CH(CCl₃)–C₆H₄–Cl

You may recognise this product as DDT, the once famous, now infamous insecticide. Its use is now controversial because of the build-up of organic chlorocompounds in animals throughout the world.

---

159. Another reactive carbonyl compound is formaldehyde. Draw out the first step of its addition to benzene in acid solution.

---

160. C₆H₆ + CH₂=⁺OH → C₆H₆(H)(⁺)–CH₂OH → C₆H₅–CH₂OH

If HCl is used as the acid catalyst, Cl⁻ instead of the aromatic ring becomes the nucleophile for the second step. Draw this.

---

161. C₆H₅–CH₂OH →(H⁺) C₆H₅–⁺CH₂ + Cl⁻ → C₆H₅–CH₂Cl

This is then a general reaction for adding a ClCH₂ group to an aromatic ring known as chloromethylation. The reaction is carried out in a single step, CH₂O and HCl being added to the aromatic compound.

REMOVAL OF CARBONYL OXYGEN, 3.13

162. The catalyst used for the DDT synthesis was H_2SO_4, that for chloromethylation is HCl. In the DDT synthesis we added two molecules of aromatic hydrocarbon, but in chloromethylation the second step had Cl^- as nucleophile. Comment?

---

163. We deliberately used H_2SO_4, an acid with a non-nucleophilic anion in the DDT synthesis, and HCl, an acid with a nucleophilic anion, in chloromethylation. We also used H_2SO_4 for the same reason in the dehydration of tertiary alcohols produced by the addition of Grignard reagents to ketones. Draw out an example of this reaction.

164.

$$\begin{array}{c} CH_3 \\ CH_3 \end{array} C=O \xrightarrow[\text{2. } H^+, H_2O]{\text{1. PhMgBr}} \begin{array}{c} CH_3 \\ CH_3 \end{array} C \begin{array}{c} OH \\ Ph \end{array} \xrightarrow{H_2SO_4}$$

$$\begin{array}{c} H \\ CH_2 \\ CH_3 \end{array} \overset{+}{C}-Ph \rightarrow \begin{array}{c} CH_2 \\ CH_3 \end{array} C-Ph$$

This then is our final example of substitution removing the carbonyl oxygen atom completely: here you see C=O being replaced by C=C.

---

165. At the end of this section, here are two review questions: aldehydes and ketones are often characterised as 2,4-dinitrophenylhydrazones, as these are usually highly crystalline orange compounds. Draw the formation of this derivative from benzaldehyde giving reagent,

REMOVAL OF CARBONYL OXYGEN, 3.14

catalyst, and the structure of the product.

166. 2,4-dinitrophenylhydrazine (O$_2$N-C$_6$H$_3$(NO$_2$)-NH·NH$_2$) + PhCHO $\xrightarrow{H^+}$ O$_2$N-C$_6$H$_3$(NO$_2$)-NH-N=CHPh

167. The smallest stable cyclic acetylene is in the nine-membered ring. If you had a sample of a nine-membered ring ketone, how would you attempt to make the cyclic acetylene from it?

168.

cyclononanone $\xrightarrow{PCl_5}$ 1,1-dichlorocyclononane $\xrightarrow{base}$ cyclononyne

This is the end of the third section of the program.

CARBANIONS AND ENOLISATION, 4.1

CONTENTS OF SECTION IV: CARBANIONS AND
                         ENOLISATION

Frame 169 Carbanions.
      172 Tautomerism.
      177 Equilibration and racemisation of ketones by enolisation.
      189 Halogenation of ketones.
      202 Formation of bromo-acid derivatives.
      208 Organo-zinc derivatives and their use in synthesis.
      213 Review questions.

CARBANIONS AND ENOLISATION, 4.2

SECTION IV: CARBANIONS AND ENOLISATION

Concepts Assumed
Hydrogen bonding.
Racemisation.
Polymerisation.

Concepts Introduced
Simple carbanions don't occur.
Stability of enolate anions.
Tautomerism.
Reactivity of enols and enolates at carbon.
Selectivity in reactions by:
  -choice of acid or base catalyst.
  -choice of metal in organo-metallic reagent.
Subtle arguments in rationalising difference between acid and base catalysed halogenation of ketones.

Concepts Reinforced
Effects of equilibria.
Drawing transition states.
Effects of substituents on stability/reactivity.
Relationship of nucleophilic addition, substitution and enolisation.
Synthesis of acid chlorides.
Use of organo-metallic reagents in synthesis.
General relation of synthesis and mechanism.

CARBANIONS AND ENOLISATION, 4.3

SECTION IV: CARBANIONS AND ENOLISATION

169. Carbonium ions are familiar intermediates in many simple reactions (S_N1, E1, etc.) and we have met many of them in this program. Carbanions are another matter. Though there are some exceptions, such as Br_3C^-, it is a good general rule to say that simple carbanions do not occur as reactive intermediates. If we want to remove a proton from a carbon atom to make an anion, we need somewhere to park the negative charge and there is nothing better for this job than the carbonyl group. Draw arrows to show the formation of an anion from acetone:

$$CH_3-C(=O)-CH_2-H \quad ^-OH \longrightarrow CH_3-C(-O^-)=CH_2$$

---

170. $$CH_3-C(=O)-CH_2-H \quad ^-OH \longrightarrow CH_3-C(-O^-)=CH_2$$

This anion, often loosely called a carbanion, is delocalised with the charge shared between the oxygen and carbon atoms. Draw arrows to show this.

---

171. $$CH_3-C(-O^-)=CH_2 \longleftrightarrow CH_3-C(=O)-CH_2^-$$

Now show how this anion can react with a proton on carbon and on oxygen to give two different products.

CARBANIONS AND ENOLISATION, 4.4

172.

$$CH_3-CO-CH_2-H \; + \; ^- \longrightarrow CH_3-CO-CH_3 \quad A$$

$$CH_3-CO-CH_2 \; {}^-\!\!\ldots H^+ \longrightarrow CH_3-C(OH)=CH_2 \quad B$$

These two compounds have identical structures except for the position of one proton: this is clearly a special case of isomerism and it is called tautomerism. A is called the keto and B the enol form of acetone.

Draw out the mechanism for the conversion of the keto to the enol form in base.

173.

$$CH_3-CO-CH_2-H \cdots {}^-OH \; \rightleftharpoons \; CH_3-CO-CH_2{}^- \cdots H-OH \; \rightleftharpoons \; CH_3-C(OH)=CH_2$$

This is an equilibrium, catalysed by acid as well as by base. Show how the enol of acetaldehyde could be formed with acid catalysis.

174.

$$CH_3-CO-H \; + \; H^+ \; \rightleftharpoons \; H_2O\cdots H-CH_2-CHOH^+ \; \rightleftharpoons \; CH_2=CH-OH$$

Now draw the enol form of this ketone: $CH_3CO.CH_2.COCH_3$.

---

175. There are in fact two possible enol forms:

$$\text{OH} \quad \text{O} \qquad\qquad \text{OH} \quad \text{O}$$

Which is the more stable?

176. The second: not only because the double bond is conjugated with the carbonyl group but because of intramolecular hydrogen bonding:

This compound in fact exists totally as the enol under normal conditions, in contrast to acetone which is entirely keto. Many other carbonyl compounds are mixtures of the two.

---

177. The 'carbanions' we formed using the carbonyl group are like enols and are called enolate anions. Draw the enolate anion from this ketone:

Ph–CO–CH=CH–CH$_3$

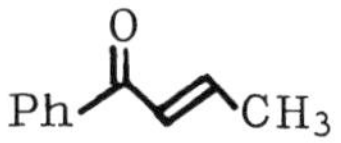

178. This time we have to remove a rather distant proton, but the charge still gets to the carbonyl group:

Draw the arrows for the re-protonation of this enolate to give the original compound.

179.

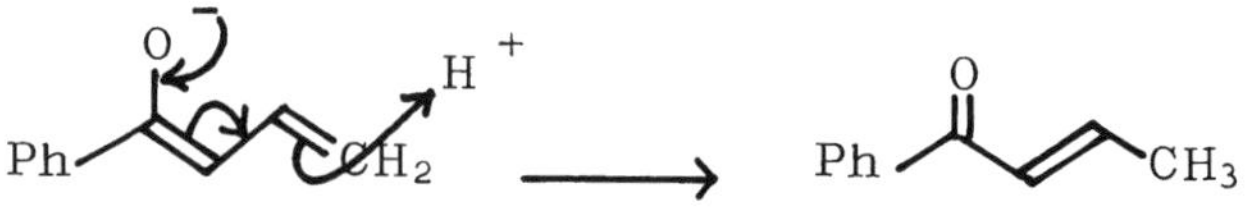

Now form the enolate anion from this ketone:

Ph–CO–CH$_2$–CH=CH$_2$ + EtO$^-$ $\xrightarrow{\text{EtOH}}$?

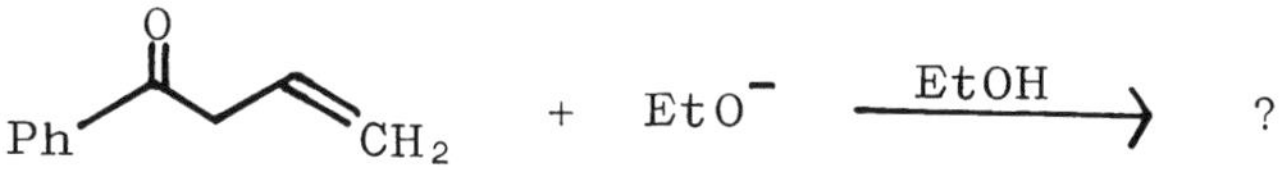

CARBANIONS AND ENOLISATION, 4.6

180. You should be adept at this by now:

[structure: Ph-C(=O)-CH(−)-CH=CH₂ with H H shown, OEt⁻] ⇌ [Ph-C(O⁻)=CH-CH=CH₂]

Now put a proton back to give a ketone.

---

181. There's a bit of a catch here. Did you notice that the enolate anions formed in frames 178 and 180 are the same? Perhaps you want to think again.

---

182. If the two enolates are the same, then they must protonate to give the same ketone. Why in fact do we form this one and not the alternative (B)?

[Ph-C(O⁻)=CH-CH=CH₂ with H-OEt] ⇌ Ph-CO-CH=CH-CH₃ (A) ; Ph-CO-CH₂-CH=CH₂ (B)

---

183. Because the whole thing is an equilibrium and the conjugated ketone (A) is the more stable. What happens then if we dissolve B in ethanol containing a small amount of EtO⁻?

---

184. Small amounts of the enolate ion will be formed which will re-protonate to give A. Ketone B is therefore quickly transformed into A.

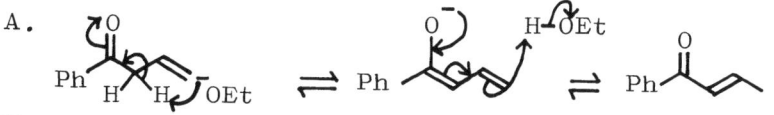

[Ph-CO-CH₂-CH=CH₂ with H H, OEt⁻] ⇌ [Ph-C(O⁻)=CH-CH=CH₂, H-OEt] ⇌ Ph-CO-CH=CH-CH₃

---

185. What happens if optically active C is dissolved in ethanol containing a catalytic amount of ethoxide ion?

[CH₃, H, Ph, COCH₃ at stereocenter] C

CARBANIONS AND ENOLISATION, 4.7

186. Again an equilibrium concentration of the enolate anion is formed, and when it is re-protonated it must give racemic product since it is planar and the proton can add on to either side:

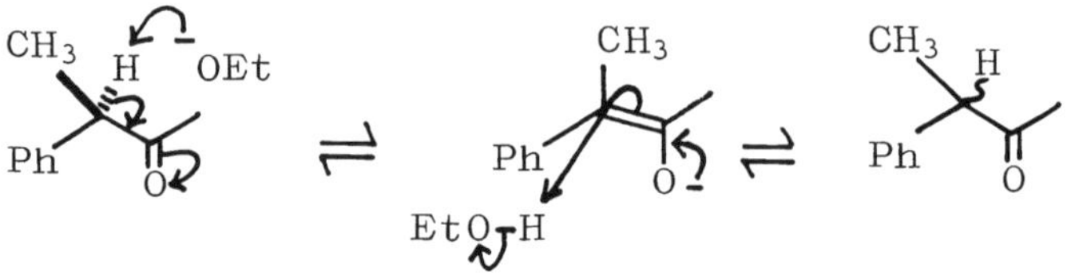

---

187. You should be getting the general idea by now: one more example of this kind of thing will do.

What happens if ketone D is dissolved under the same conditions?

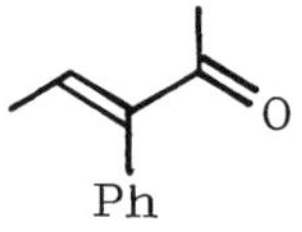

---

188. This time the enolate anion can rotate about what was a double bond in the original compound, giving a mixture of cis and trans forms.

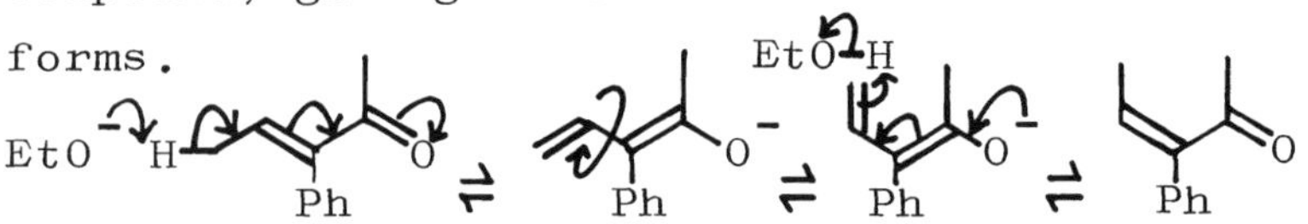

This is a good place to stop if you want a rest.

189. So far we have reacted enolate ions with protons. They also react with many electrophiles in just the same way: the electrophile attacking at carbon. Make the enolate ion from PhCO.Me and attack it with bromine.

190. If you're in trouble, bring the negative charge back from the oxygen atom, just as you have been doing so far, but this time attack one end of the bromine molecule with it expelling the other bromine atom as Br⁻.

191.

$$Ph-\underset{\underset{H}{\parallel}}{C}-CH_2 \cdot \cdot \cdot OH \rightleftharpoons Ph-\underset{\underset{}{\parallel}}{C}=CH_2 \cdot \cdot \cdot Br-Br \longrightarrow Ph-\underset{\underset{}{\parallel}}{C}-CH_2Br$$

With base still present in the reaction mixture, would anything happen to this product?

192. Bromide ion could be displaced, but in fact it isn't. Have you noticed that there are two more protons waiting to be removed?

193. The process is repeated with another proton:

$$Ph-\underset{\underset{H}{\parallel}}{C}-\underset{\underset{H}{|}}{C}HBr \cdot \cdot \cdot OH \rightleftharpoons Ph-\underset{\underset{}{\parallel}}{C}=CHBr \cdot \cdot \cdot Br-Br \longrightarrow Ph-\underset{\underset{}{\parallel}}{C}-CHBr_2$$
                                                                A

Will this second enolate ion (A) be formed more or less easily than the first (in frame 191)?

194. <u>More</u> easily, because the inductive effect of the bromine atom makes the hydrogen easier to remove. We can't therefore stop this reaction at the mono-bromo stage, nor at the di-bromo stage: what happens next?

CARBANIONS AND ENOLISATION, 4.9

195. The process is repeated for a third time:

Ph-C(=O)-CHBr-Br + ⁻OEt ⇌ Ph-C(-O⁻)(Br)-CBr₂···Br → Ph-C(=O)-CBr₃

So in the presence of enough bromine, the ketone is converted rapidly into the <u>tri</u>-bromo ketone. But the reaction doesn't stop, even here. Do you remember that we singled out CBr₃⁻ as one of the rare stable simple carbanions? How could another molecule of hydroxide ion attack the product to give this anion?

---

196. In a straightforward substitution reaction:

Ph-C(=O)-CBr₃ + HO⁻ ⇌ Ph-C(O⁻)(OH)-CBr₃ → PhCO₂H + ⁻CBr₃

The carbanion, though <u>relatively</u> stable, compared to other carbanions, is still reactive enough to pick up a proton from water to give CHBr₃.

Write down a summary equation for this whole process, starting from PhCO.Me.

---

197. PhCOMe + Br₂ + HO⁻ ⟶ PhCO₂H + CHBr₃

This reaction works equally well with chlorine or iodine and is known as the chloroform or iodoform reactions in these cases after the names of the products.

CARBANIONS AND ENOLISATION, 4.10

198. We do, of course, need a way of making the <u>mono</u> bromo derivative of a ketone, and it turns out that we can do this by halogenation in acid solution. Take the same ketone, PhCOMe, make the enol in acid solution, and react it with bromine.

---

199. Congratulations if you got this right all in one go!

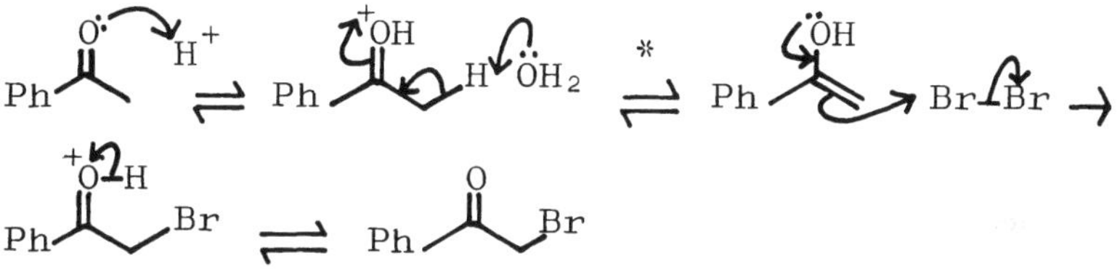

Ph-CO-Me + H⁺ ⇌ Ph-C(⁺OH)=CH₂ ··· H-OH₂ ⇌ Ph-C(OH)=CH₂ + Br-Br → Ph-C(⁺OH)(H)-CH₂Br ⇌ Ph-CO-CH₂Br

---

200. Now we come to the slightly tricky question of why this reaction can be stopped at this stage. Let's look at the enolisation step (marked * in frame 199). Draw the transition state for this step. If you've forgotten how to draw transition states look at frames 54 – 58.

201.

[Two transition state structures shown side by side: the left one without Br, the right one with Br substituent, both showing (+)OH···H···OH₂(+) partial bonds]

We want to compare this transition state with the one for the same step with the mono bromo compound, so I have drawn them side by side. Note that in these transition states a positive charge is smeared out over all the atoms connected by the dotted line. Therefore, any substituent like Br which is electron withdrawing will destabilise the transition state. So the

CARBANIONS AND ENOLISATION, 4.11

first step to give the mono bromo compound goes faster than the step to give the dibromo compound and we can stop the reaction by using only so much bromine at the first stage.

---

202. We have made acid chlorides in the program already (frames 103-106) by the reaction between $SOCl_2$ and $RCOOH$. PCl_5 and PBr_5 do much the same thing:

$$CH_3CO_2H + PBr_5 \longrightarrow CH_3CO.Br + POBr_3 + HBr$$

What further reaction might happen under these conditions if bromine were also present in the reaction mixture?

---

203. A clue is in the presence of HBr, which could catalyse the enolisation of the acid bromide.

---

204. The acid bromine enolises and reacts with bromine:

$$CH_3-\overset{O}{\underset{\|}{C}}-Br \;\overset{H^+}{\rightleftharpoons}\; Br^- \;\;H-CH_2-\overset{HO^+}{\underset{\|}{C}}-Br \;\rightleftharpoons\; Br \overset{O^-}{\underset{}{\diagup}} \;Br\text{-}Br$$

$$Br-\overset{O}{\underset{\|}{C}}-Br$$

Since this is an acid-catalysed reaction it can be stopped at this stage. This valuable reaction, known as the Hell-Volhard-Zelinsky (HVZ) reaction, is normally carried out by treating the acid with red phosphorus and bromine. PBr_5 is thus made on the spot.

---

205. What would be the product in this reaction? $PhCH_2CO_2H \xrightarrow{\text{red P} + Br_2}$

CARBANIONS AND ENOLISATION, 4.12

206. PhCH$_2$CO$_2$H $\longrightarrow$ PhCHBr.COBr

And what in this reaction?

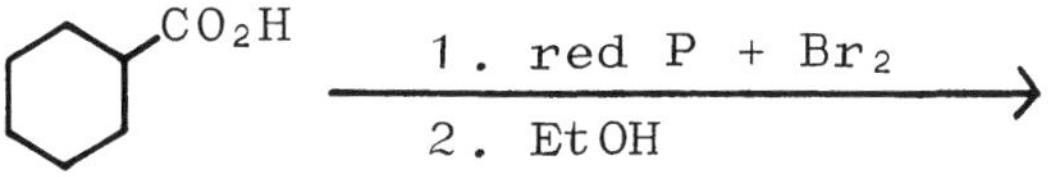

cyclohexyl-CO$_2$H $\xrightarrow{\text{1. red P + Br}_2}{\text{2. EtOH}}$

207.

cyclohexyl-CO$_2$H $\xrightarrow[\text{Br}_2]{\text{red P}}$ 1-bromocyclohexyl-COBr $\xrightarrow{\text{EtOH}}$ 1-bromocyclohexyl-CO$_2$Et

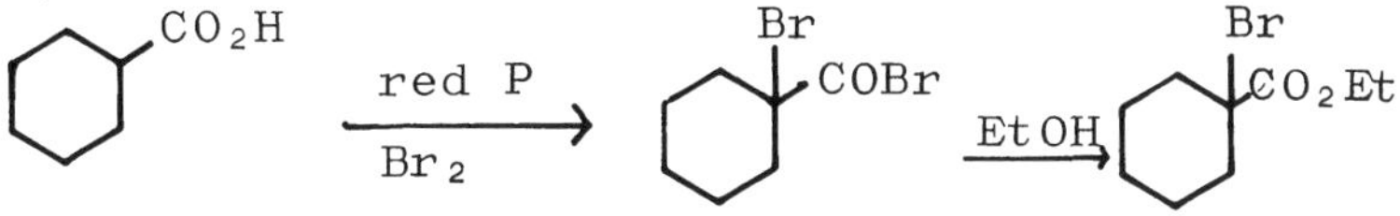

We don't usually try to make Grignard reagents from α-bromo esters like this compound. Why not?

---

208. Because Grignard reagents react with esters, and the Grignard reagent from this compound would therefore react with itself, that is polymerise.

We use the zinc compound instead as these are less reactive and don't polymerise. We make them in the same way as Grignard reagents, and they react in the same way with ketones. What would happen here:

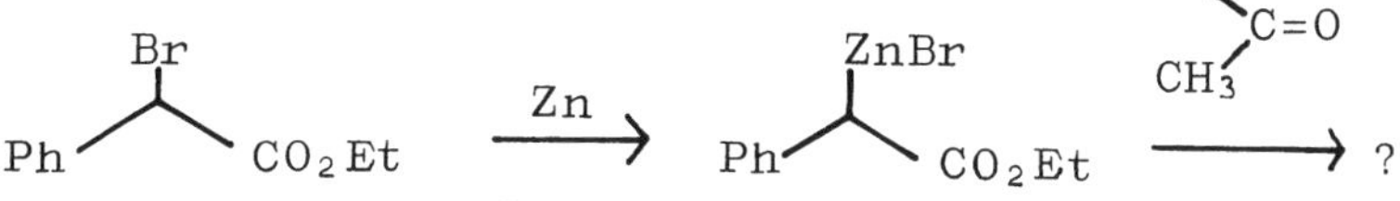

Ph(Br)C(CH$_3$)CO$_2$Et $\xrightarrow{\text{Zn}}$ Ph(ZnBr)C(CH$_3$)CO$_2$Et $\xrightarrow{\text{(CH}_3\text{)}_2\text{C=O}}$?

209.

BrZn–C(Ph)(CH$_3$)CO$_2$Et + (CH$_3$)$_2$C=O $\longrightarrow$ (CH$_3$)$_2$C(O$^-$)–C(Ph)(CH$_3$)CO$_2$Et $\xrightarrow{\text{H}^+}$ (CH$_3$)$_2$C(OH)–C(Ph)(CH$_3$)CO$_2$Et

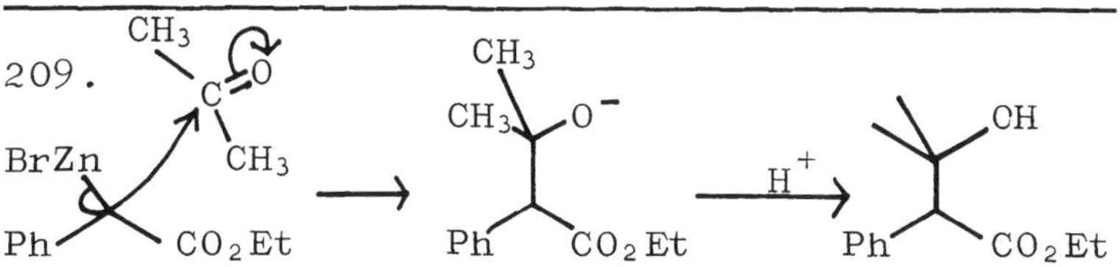

What do you think happens to this during the acidic work up?

CARBANIONS AND ENOLISATION, 4.13

210. It dehydrates spontaneously to give an α-β unsaturated ester. We have already met this kind of reaction in frame 40.

Ph–C(CH₃)(OH₂⁺)–CO₂Et ⟶ Ph–C(CH₃)(H)–CO₂Et ⟶ PhC(CH₃)=C(CH₃)–CO₂Et

211. We said that these reactions are valuable so let's prove it. How would you carry out this synthesis:

PhCOCH₃ ⟶ PhCMe=CMe.CO₂Et

212.
CH₃CH₂COOH $\xrightarrow{HVZ}$ CH₃CHBr.COBr $\xrightarrow{EtOH}$ CH₃CHBr.CO₂Et

PhCOCH₃ + CH₃CH(ZnBr).CO₂Et ⟶ PhC(OH)(CH₃)–CH(CH₃).CO₂Et

$\xrightarrow{H^+}$ PhCMe=CMe.CO₂Et

213. In the next section we shall be looking at more reactions of enols which make new carbon-carbon bonds. We'll finish off this section with review questions. Draw the most stable enol from:

cyclohexane-1,3-dione ; CH₃COCH₂COOEt

214. (enol of cyclohexane-1,3-dione with OH)   (H-bonded enol of acetoacetate, two resonance forms shown)

CARBANIONS AND ENOLISATION, 4.14

215. Arrange these compounds in order of reactivity towards i) nucleophilic addition, ii) nucleophilic substitution, and iii) enolisation:

CH_3COOEt CH_3COCl CH_3CONH_2 $CH_3COO.COCH_3$

216. The order is the same in each case:
  most reactive: acid chloride
           anhydride
           ester
  least reactive: amide

because each reaction involves negative charge being taken on the carbonyl oxygen atom:

addition: substitution: enolisation:

$$\underset{Y^-}{\overset{\displaystyle CH_3 \overset{O^-}{\underset{|}{C}} X}{}} \qquad \underset{Y^-}{\overset{\displaystyle CH_3 \overset{O}{\underset{|}{C}} X}{}} \qquad \underset{Y^-}{\overset{\displaystyle H-CH_2 \overset{O^-}{\underset{|}{C}} X}{}}$$

If you are in doubt about this, ask your adviser. This is the end of the fourth section.

BUILDING ORGANIC MOLECULES, 5:1

CONTENTS OF SECTION V: BUILDING ORGANIC MOLECULES FROM CARBONYL COMPOUNDS

Frame 217 Using enols as nucleophiles to attack other carbonyl groups. The aldol reaction.
- 221 The Claisen ester condensation.
- 227 Acid catalysed condensation of acetone.
- 230 Self condensation reactions.
- 233 Elaboration of a skeleton in synthesis.
- 235 Cross condensations with molecules which cannot enolise.
- 248 Mannich reaction.
- 252 Perkin reaction.
- 260 Stable enols from β-dicarbonyl compounds.
- 268 Knoevenagel reaction.
- 275 Alkylation of β-dicarbonyl compounds.
- 281 Michael reaction.
- 295 Decarboxylation.
- 302 Base cleavage of β-dicarbonyl compounds.
- 307 Cyclisation reactions: the Dieckmann condensation.
- 310 Cyclisation of diketones.
- 312 The dimedone synthesis.
- 316 Ring opening by base cleavage of β-dicarbonyl compounds.
- 318 Revision questions.
- 324 Examples of syntheses: part of two steroid syntheses.

BUILDING ORGANIC MOLECULES, 5:2

CONTENTS CONTINUED

Frame 328   Stork's cedrene synthesis.

BUILDING ORGANIC MOLECULES, 5:3

SECTION V: BUILDING ORGANIC MOLECULES FROM CARBONYL COMPOUNDS

Concepts Assumed:

Relative reactivities of aldehydes, ketones, esters, acid chlorides and anhydrides.

Concepts Introduced:

Carbon-carbon bond formation from two components: enol and electrophile.

Condensation reactions catalysed by base or acid.

Selection of right base for condensation reactions.

Ambiguity of condensation reactions.

Self condensations and cross condensations.

Use of compounds with no enolisable protons.

Methods of avoiding ambiguity in cross condensations.

Secondary amines as catalysts via $R_2C=NR_2^+$.

Stability of five and six membered rings.

Instability of four membered rings.

Use of β-dicarbonyl compounds to form stable enolates.

Use of NO_2 and CN as carbonyl equivalents.

Nucleophilic addition to enones.

Decarboxylation and β-dicarbonyl cleavage reactions.

Use of carbonyl groups as activating and directing groups in syntheses.

Need to plan ahead in syntheses.

Concepts Reinforced:

Basic carbonyl mechanisms.

Use of pK_as.

BUILDING ORGANIC MOLECULES, 5:4

Concepts Reinforced continued:
Elimination of OH and R_2N in acid conditions.
Similarity of reactions of acyclic and cyclic compounds.

BUILDING ORGANIC MOLECULES, 5:5

SECTION V: BUILDING ORGANIC MOLECULES FROM
                     CARBONYL COMPOUNDS

217. In this part of the program we are going to explore ways of making organic compounds from carbonyl compounds. The reactions of carbanions and enols in particular are some of the chief ways in which the organic chemist builds up complicated structures. Many people find this a daunting part of organic chemistry and so I have treated the subject in rather more detail than other subjects in the program. If you are reading this you must have survived the earlier sections with some enjoyment and you ought easily to be capable of tackling this rather more demanding section.

218. Let's begin by stating very clearly just what possibilities we are considering. We have discussed <u>four</u> different types of reaction which can occur between a nucleophile (base) and a carbonyl compound. List these in words.

219. They are: a) Addition
                b) Substitution
                c) Complete removal of carbonyl oxygen.
                d) Removal of an α-proton: enolisation.

One possibility we have not yet considered is combining d) with the other three: that is using the enolate anion as the nucleophile in these reactions.

BUILDING ORGANIC MOLECULES, 5:6

Make the enolate anion from acetaldehyde and add it to the carbonyl group of another molecule of acetaldehyde.

220.

$$H-\overset{O}{\underset{}{C}}-CH_3 \rightleftharpoons H-\overset{O^-}{\underset{}{C}}=CH_2 \quad \underset{CH_3}{\overset{H}{\underset{|}{C}=O}} \longrightarrow$$

$$H-\overset{O}{\underset{}{C}}-CH_2-\overset{O^-}{\underset{}{CH}}-CH_3 \longrightarrow H-\overset{O}{\underset{}{C}}-CH_2-\overset{OH}{\underset{}{CH}}-CH_3$$

In making a four carbon chain from two two-carbon units we used a nucleophilic enolate and an electrophilic carbonyl group. This theme of the two components is universal to this section.

221. Now make the enolate from ethyl acetate and attack another molecule of ethyl acetate with it.

222. Have you just added the one to the other, or have you extended the reaction, as you should, to make it a substitution?

223. Here then is the full reaction:

$$EtO-\overset{O}{\underset{}{C}}-CH_3 \rightleftharpoons EtO-\overset{O^-}{\underset{}{C}}=CH_2 \quad \underset{CH_3}{\overset{OEt}{\underset{|}{C}=O}} \longrightarrow$$

$$EtO-\overset{O}{\underset{}{C}}-CH_2-\underset{CH_3}{\overset{OEt}{\underset{|}{C}}-O^-} \longrightarrow EtO-\overset{O}{\underset{}{C}}-CH_2-\overset{O}{\underset{}{C}}-CH_3$$

224. These reactions were catalysed by an unspecified base. With the aldehyde, quite weak base is needed and there is no possibility of an alternative reaction, but we couldn't use hydroxide ion, for example, to enolise the ester. Why not?

---

225. Hydroxide ion would attack the carbonyl group and hydrolyse the ester.
We have an ingenious way of getting round this problem. There is one base, and one only, which we can use to enolise ethyl acetate so that even if it <u>does</u> attack the carbonyl group and displace EtO$^-$, we don't care. What base is that?

---

226. Clearly EtO$^-$!
If it attacks the carbonyl group it simply makes another molecule of ethyl acetate and we don't notice:

$$EtO^- \overset{\overset{\displaystyle C\!-\!OEt}{\big|}}{\underset{\underset{\displaystyle CH_3}{\big|}}{C}} \longrightarrow \underset{O}{\overset{EtO}{>}}\!C\!-\!CH_3 + EtO^-$$

---

227. These reactions can also be catalysed by acid. Try your hand at enolising acetone in acid solution and then using the enol to attack another molecule of acetone, still in acid.

228.

$$CH_3-\overset{+\overset{\cdot}{O}H}{\underset{\|}{C}}-CH_2-H \rightleftharpoons CH_3-\overset{\cdot\overset{\cdot}{O}H}{\underset{}{C}}=CH_2 \overset{}{\underset{}{\longrightarrow}} \overset{CH_3}{\underset{CH_3}{\overset{|}{C}=O}}\overset{+}{H} \longrightarrow$$

$$CH_3-\overset{+\overset{}{O}H}{\underset{\|}{C}}-CH_2-\overset{OH}{\underset{\overset{|}{CH_3}}{\overset{|}{C}}}-CH_3 \overset{-H^+}{\longrightarrow} CH_3-\overset{O}{\underset{\|}{C}}-CH_2-\overset{OH}{\underset{\overset{|}{CH_3}}{\overset{|}{C}}}-CH_3$$

You may have carried this reaction a step further. If you didn't, consider now what would happen to that product under the reaction conditions.

---

229. The tertiary alcohol will be protonated and lost:

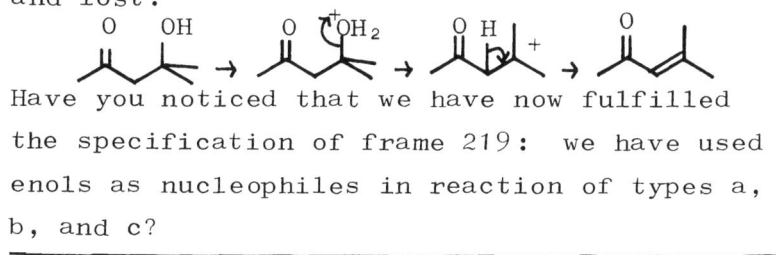

(structures showing O-OH → O-⁺OH₂ → O H⁺ → enone)

Have you noticed that we have now fulfilled the specification of frame 219: we have used enols as nucleophiles in reaction of types a, b, and c?

---

230. Reactions of this type in which carbonyl compounds combine together to form longer carbon chains are (loosely) known as condensation reactions. So far we have considered only <u>self-condensations</u>, that is ones in which both the nucleophilic component and the electrophilic component are forms of the same molecule. How would you make this compound:

Ph–CH(CO₂Et)–C(=O)–CH₂–Ph

231. By self condensation of an ester, using EtO⁻ as base:

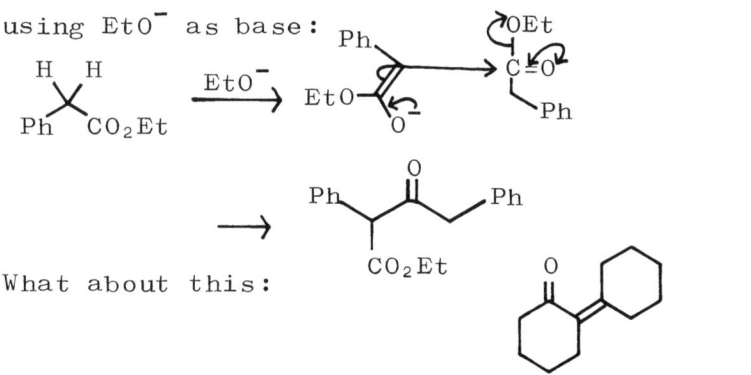

[mechanism scheme: PhCH₂CO₂Et + EtO⁻ → enolate → attack on second ester → PhCH(CO₂Et)-C(=O)-CH₂Ph]

→ Ph–CH(CO₂Et)–C(=O)–CH₂Ph

What about this: [cyclohexanone with =cyclohexylidene substituent]

---

232. By self condensation of cyclohexanone, using acid or base as catalyst:

[mechanism: cyclohexanone →(H⁺) enol, attacks protonated cyclohexanone → 1-(1-hydroxycyclohexyl)cyclohexan-2-one → (–H₂O) → 2-cyclohexylidenecyclohexanone]

---

233. Small structural changes can easily be introduced after the basic skeleton is complete, a process often known as elaboration. How would you synthesise this compound:

Ph–CH(CH₂OH)–CH(OH)–CH₂–Ph

---

234. The basic skeleton is the same as the one we made in frame 230. All we have to do is to reduce both ester and ketone, and the reagent for this is LiAlH₄ (frames 43-5 and 78-9).

---

235. So far we have used the same molecule to provide both nucleophilic (enolate) and electrophilic (carbonyl) components. This

avoids ambiguity, but we must also be able to do cross-condensations in which different molecules supply the different components. What problem would arise if we tried to do this condensation:

enolate component: carbonyl component:

$$CH_3CH(OEt)=O + CH_3C(OEt) \xrightarrow{EtO^-} EtO_2C-CH(CH_3)-C(CH_3)=O$$

236. Though we have labelled them, the molecules can't be expected to know which role they're supposed to fulfil and may give products of self-condensation.

$$CH_3CH_2CO_2Et \longrightarrow EtO_2C-CH(CH_3)-CO.CH_2CH_3$$
$$CH_3CO_2Et \longrightarrow EtO_2C.CH_2COCH_3$$

This is a good place to rest if you want to.

237. One way round this problem is to use components which cannot self-condense. Select from this list some molecules which fit the bill:

a) t-BuCHO   b) MeCOOBu-t   c) PhCHO
d) EtOCO.OEt e) $CH_2=O$ f) HCOOEt
g) PhCH$_2$CHO h) EtO.CO.CO.OEt

238. All except b and g cannot self condense because they have no enolisable protons. These are very useful compounds.
Make a classification of the type of substituent we can put on a carbonyl group so that it

can't self condense. My list has five types.

239. Here is my list, yours may be different:
i) tertiary alkyl groups like t-Bu.
ii) aryl groups like Ph.
iii) hydrogen, H
iv) electronegative substituents like Cl, OEt.
v) other carbonyl groups as in 237h.
With a list like this we can invent molecules which can't enolise. Write down a ketone and an acid chloride which can't enolise.

240. There are lots of answers such as:
Ketone: PhCOPh PhCOBu-t
Acid chloride: ClCOOEt PhCOCl Me$_2$N.COCl
One of these compounds can act only as the electrophilic component in a condensation, and can't self condense. What will be formed here:

O_2N-C$_6$H$_4$-CHO + PhCOCH$_3$ $\xrightarrow{H^+}$

241. The only compound which can enolise is the methyl ketone:

Ph-CO-CH$_3$ $\underset{}{\overset{H^+}{\rightleftharpoons}}$ Ph-C(OH)=CH$_2$

and it will then attack the aldehyde, which can only act as the electrophilic component:

Ph-C(OH)=CH$_2$ + CH(=O)-C$_6$H$_4$-NO$_2$ → Ph-CO-CH$_2$-CH(OH)-C$_6$H$_4$-NO$_2$

$\xrightarrow[-H_2O]{H^+}$ Ph-CO-CH=CH-C$_6$H$_4$-NO$_2$

BUILDING ORGANIC MOLECULES, 5:12

242. There was in fact one ambiguity left in this last example. The aldehyde could act only as the electrophilic component, but the ketone could act as both components, that is, it could self condense. In fact it didn't. Why not?

243. If you don't see this at once, imagine what happens when a molecule of ketone enolises. It then begins to look around for an electrophilic molecule, and it has two choices: another molecule of itself, or the aldehyde...

244. The point here is that the aldehyde is more reactive towards nucleophiles than the ketone: so, when the ketone enolises, it will always attack the more reactive aldehyde, and never the less reactive ketone. Aldehydes, compounds with two adjacent carbonyl groups, and formate esters (HCOOR) are suitable reactive compounds for this type of condensation. What would be formed here:

cyclohexanone + HCOOEt $\xrightarrow{\text{base}}$?

245. The ketone alone can enolise, and the enol will prefer to attack the formate ester:

246. So here we have our first safe method of doing a cross condensation. We use any enolisable compound as the nucleophilic component, but we must have as the electrophilic component...what?

---

247. A compound which can't enolise, but which is more reactive than the other compound towards nucleophiles.

---

248. One useful compound for this type of reaction is formaldehyde, $CH_2=O$, and we often carry out condensations with this using a secondary amine and dilute acid. The amine gives this reaction:

$$CH_2O + R_2NH \xrightarrow{H^+} CH_2=\overset{+}{N}R_2$$

Draw a mechanism for this.

---

249. $R_2\ddot{N}H \curvearrowright CH_2=\overset{\frown}{O} \xrightarrow{H^+} R_2\overset{+}{\underset{H}{N}}-CH_2-OH$

$\rightleftharpoons R_2\ddot{N}-CH_2-\overset{+}{O}H_2 \longrightarrow R_2\overset{+}{N}=CH_2$

This reaction is very like the condensation of primary amines with ketones (frames 123-5). If we have PhCOMe present in the solution, what will happen now?

---

250. $CH_2=\overset{+}{N}R_2$ is of course very electrophilic and picks up the enol from the ketone:

$$Ph\overset{O}{\underset{}{\overset{\|}{C}}}CH_3 \xrightarrow{H^+} Ph\overset{\overset{\cdot\cdot OH}{}}{=}CH_2 \curvearrowright CH_2=\overset{+}{N}R_2 \rightarrow Ph\overset{O}{\underset{}{\overset{\|}{C}}}CH_2CH_2NR_2$$

This is known as the Mannich reaction. What product would be formed here:

$$\text{[bicyclic ketone]} \xrightarrow[H^+]{CH_2O, \, Me_2NH} \;?$$

251.

[Structure: enol formation from 2-methyl-tetralone with H⁺, then reaction with CH₂=NMe₂⁺ (Mannich reagent) giving the α-alkylated product with CH₂CH₂NMe₂ side chain]

252. A kind of enolic component we haven't mentioned yet is the acid anhydride. If you wanted to make the enolate anion from acetic anhydride, what base would you recommend?

253. We would use acetate ion as base, since any substitution reaction then simply regenerates the anhydride.
Using the anhydride then as the enolic component, what happens with PhCHO as the electrophilic component?

254.

[Mechanism: acetate deprotonates acetic anhydride; enolate attacks PhCHO to give alkoxide intermediate CH₃C(O)OC(O)CH₂CH(O⁻)Ph]

The O⁻ is ideally placed to attack one of the carbonyl groups in the molecule intramolecularly. Can you draw this?

255. It could attack the nearer carbonyl group making a four membered ring:

[Scheme showing four-membered ring formation crossed out (disfavoured) vs six-membered ring]

or the other making a six membered ring. As you know (frames 59ff.) six membered rings are very stable, and this is the preferred reaction:

[Reaction scheme showing equilibrium and arrow with three structures involving OAc, Ph, and O groups]

256. Under the conditions of the reaction, an elimination occurs:

[Scheme showing elimination reaction producing cinnamate-type intermediate and then HO₂C-CH=CH-Ph]

and acid work up gives as a final product an α-β unsaturated acid. How would you make this compound?

$$O_2N\text{-C}_6H_4\text{-CH=C(CH}_3\text{)-CO}_2H$$

257.

O_2N-C₆H₄-CHO + $(CH_3CH_2CO)_2O$ + $CH_3CH_2CO_2^-$

$\longrightarrow$ product

Note that with the change of acid anhydride, a change of base is called for.

---

258. Just to check that you've understood these ideas, what would happen if we tried to condense acetaldehyde as enolic component with benzophenone ($Ph_2C=O$)?

---

259. Acetaldehyde would enolise all right, but it would never react with $Ph_2C=O$. It would much prefer to react with another molecule of itself:

[Mechanism scheme:]

$$H\text{-CO-CH}_3 \rightleftharpoons H\text{-C(O}^-\text{)=CH}_2 \quad \overset{O}{\underset{}{\|}}{\text{CH-CH}_3} \rightarrow H\text{-CO-CH}_2\text{-CH(OH)-CH}_3$$

BUILDING ORGANIC MOLECULES, 5:16

This is a good place to rest if you want to.

260. We have done a number of cross condensations so far by using a reactive electrophilic component which doesn't act as a nucleophile because it can't enolise. Now we're going to look at the reverse: a reactive enolic component which doesn't act as an electrophile. We thus want to increase the ability of a compound to enolise without making the carbonyl group more electrophilic. One way to do this is to use <u>two</u> carbonyl groups, both of which encourage the enolisation of the <u>same</u> CH group. Using ester functions, draw such a compound.

---

261. The simplest compound of this sort is:

$$EtO^- \quad H-CH\begin{smallmatrix}C-OEt\\C-OEt\end{smallmatrix} \rightarrow CH\begin{smallmatrix}C-OEt\\C-OEt\end{smallmatrix} \leftrightarrow CH\begin{smallmatrix}C-OEt\\C-OEt\end{smallmatrix}$$

261a

What base would we use to make the enolate anion here?

---

262. Ethoxide ion, so that it doesn't matter if substitution occurs. We shall get a substantial amount of enolate here because the pK$_a$ of EtOH is about 16, and that of the CH's in 261a is about 14.

Select which of the following compounds would form substantial amounts of enolate anion with ethoxide ion in ethanol.

BUILDING ORGANIC MOLECULES, 5:17

List of compounds:
a. PhCOMe   b. CH_3NO_2
c. CH_3COCH_2COOEt d. H.CHO
e. t-BuCHO   f. $CH_2(COOEt)_2$
g. $EtO.CO.CH_2CN$ h. $(COOEt)_2$
i. $PhCH_2CHO$ j. $PhCOCH_2COCH_3$

---

263. b,c,f,g, and j are reactive enough. If you got all these and included no others, skip to frame 268. Otherwise read on: if you included a or i read frame 264; if you included d,e, or h, read 265; if you omitted b, read 266; if you omitted c,f,g, or j, read 267.

---

264. To get a substantial amount of enolate anion, the pK_a of the carbonyl compound wants to be several pH units below that of the alkoxide ion. In fact the limit is about 14 using ethoxide ion. The compounds you included are therefore too weakly acidic with pK_as of about 20(a) and 18(i). Go to frame 268.

---

265. Yes, these compounds are very reactive in the electrophilic sense, but they have no enolisable hydrogen atoms! Go to frame 268.

---

266. It is rather surprising to find a compound with only one activating group among the others, but the nitro group is very electron-withdrawing so that the pK_a of nitromethane is 10: it is even more acidic than most dicarbonyl compounds! Go to frame 268.

BUILDING ORGANIC MOLECULES, 5:18

267. The chief group of compounds acidic enough to provide substantial amounts of enolate ion with ethoxide base is the β-dicarbonyl compounds. They have pK_as in the region 9-13.

---

268. Draw the enolate anion from cyanoacetic ester $EtOOC.CH_2.CN$, and show how it is stable.

269.

$$EtO-\overset{O}{\underset{\|}{C}}-\overset{H}{\underset{}{CH}}-C\equiv N \xrightarrow{OEt^-} EtO-\overset{\bar{O}}{\underset{\|}{C}}-CH-C\equiv N \leftrightarrow EtO-\overset{O}{\underset{\|}{C}}-CH=C=N^-$$

Now use this anion to attack a molecule of acetone.

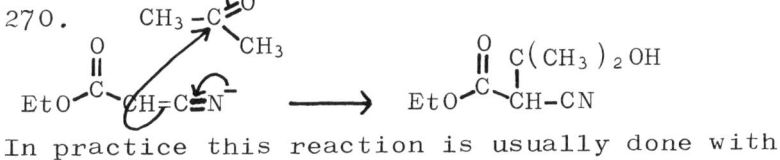

270.

$$EtO-\overset{O}{\underset{\|}{C}}-CH=C\equiv N^- \quad + \quad CH_3-\overset{O}{\underset{}{C}}-CH_3 \longrightarrow EtO-\overset{O}{\underset{\|}{C}}-\overset{C(CH_3)_2OH}{\underset{}{CH}}-CN$$

In practice this reaction is usually done with catalysis by a combination of a secondary amine and acid. Can you suggest a role for the secondary amine?

---

271. It combines with the electrophilic component (acetone in this case) to make it even more electrophilic. We met this previously in the Mannich reaction (frames 248-250).

$$(CH_3)_2CO + R_2NH \xrightarrow{H^+} (CH_3)_2\overset{\overset{+}{N}R_2}{\underset{}{C}} \xleftarrow{} CH=C\overset{OEt}{\underset{OH}{\diagdown}}$$
$$ \qquad\qquad\qquad\qquad\qquad CN$$

$$\longrightarrow \quad EtO_2C-\overset{Me_2C-NR_2}{\underset{}{CH}}-CN$$

Under these acidic conditions, a further reaction will take place. Can you suggest what?

272. If you don't know, look at the product and see where protonation is most likely to occur: one of the groups is protonated and then falls off.

---

273. An elimination reaction occurs:

Me$_2$C-NR$_2$ / EtO$_2$C-CH-CN →(H$^+$) Me$_2$C-NHR$_2^+$ / EtO$_2$C-CH-CN (with H) → Me$_2$C=C(EtO$_2$C)(CN)

This reaction goes under the name of the Knoevenagel reaction.

How would you make this compound:

PhCH=C(COCH$_3$)(CO$_2$Et) ?

---

274. By a Knoevenagel reaction between benzaldehyde and ethyl acetoacetate:

PhCHO + CH$_2$(COCH$_3$)(CO$_2$Et) —Me$_2$NH / H$^+$→ PhCH=C(COCH$_3$)(CO$_2$Et)

---

275. The enolates from β-dicarbonyl compounds are so easily formed that they can be used in a very simple carbon-carbon bond forming reaction outside our general scheme. Consider what would happen if you made the enolate anion from the compound below and reacted it with methyl iodide.

[2-(ethoxycarbonyl)cyclopentanone]   1. EtO$^-$ 2. MeI   →   ?

276. The enolate anion attacks methyl iodide in an S_N2 reaction, in other words it is alkylated:

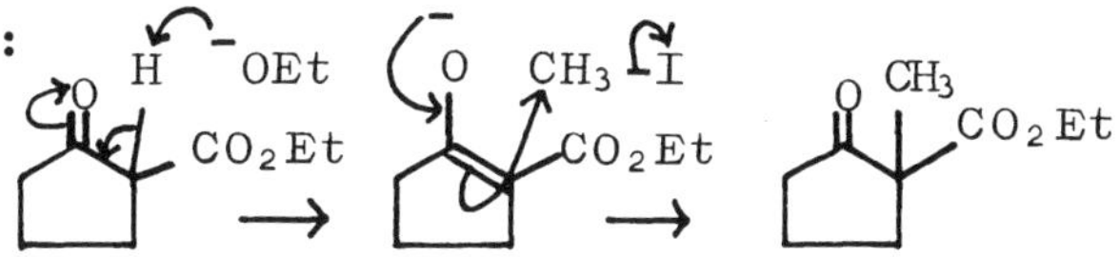

[reaction scheme showing enolate of ethyl 2-oxocyclohexanecarboxylate attacking CH₃I to give the α-methylated product]

This reaction extends the scope of our work with β-dicarbonyl compounds considerably. Do you remember how we synthesised them in the program? How, for example, would you make this:

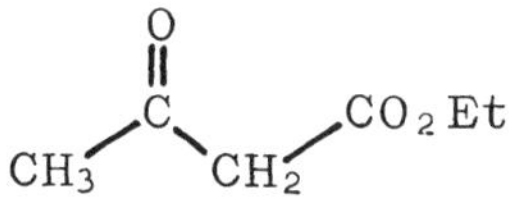

CH₃−CO−CH₂−CO₂Et

277. By a self-condensation reaction of an ester:

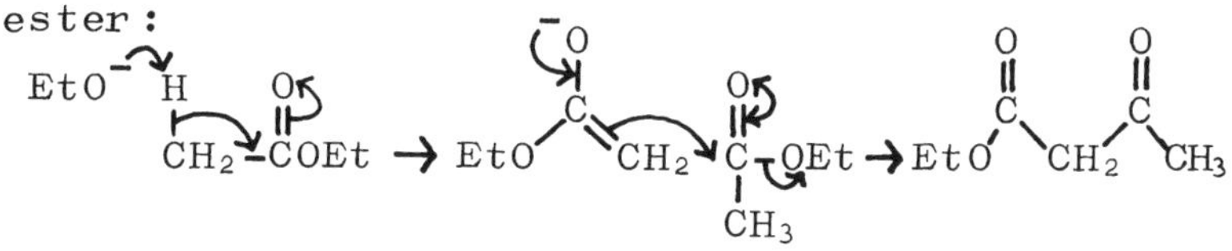

EtO−H CH₂−COEt → EtO−CO−CH₂ + EtO−C(CH₃)(O⁻)−OEt → EtO−CO−CH₂−CO−CH₃

How then would you make this:

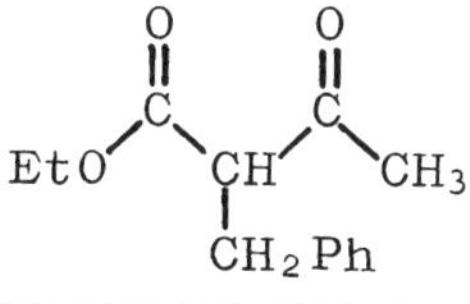

EtO−CO−CH(CH₂Ph)−CO−CH₃

278. First make the right β-dicarbonyl compound and then alkylate it with the right alkyl halide:

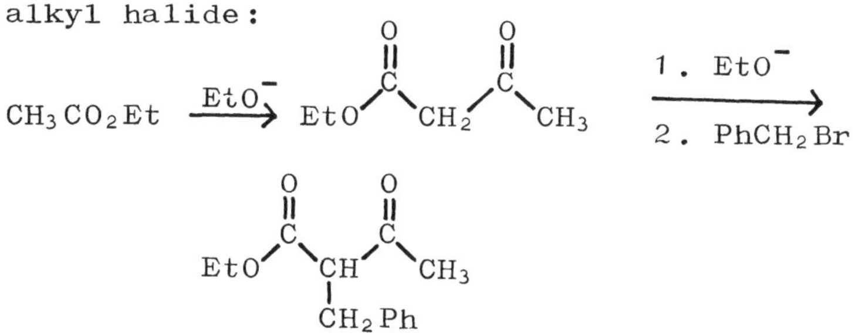

$CH_3CO_2Et \xrightarrow{EtO^-}$ EtO−CO−CH₂−CO−CH₃ $\xrightarrow[2.\ PhCH_2Br]{1.\ EtO^-}$

EtO−CO−CH(CH₂Ph)−CO−CH₃

279. We might summarise our syntheses so far in terms of the general classes of molecules we are able to make. Look back over the program (section 5) and see if you can make a short list.

---

280. We can make:

i) β-hydroxy carbonyl compounds by addition

$$CH_3CHO \xrightarrow{base} CH_3-\underset{\underset{H}{|}}{\overset{OH}{\underset{|}{C}}}-CH_2-CHO$$

ii) α-β-unsaturated carbonyl compounds by loss of water from the β-hydroxy compounds:

$$(CH_3)_2C=O + PhCHO \longrightarrow PhCH=CH-\overset{O}{\overset{\|}{C}}CH_3$$

iii) β-dicarbonyl compounds by substitution:

[cyclohexanone] + HCO₂Et $\xrightarrow{EtO^-}$ [2-formylcyclohexanone]

Have you noticed that in all these compounds there is the same relationship between the two original carbonyl groups? The one supporting the enol and the electrophilic one always end up in a 1,3-relationship. Mark these two carbon atoms (*) on each of the products in this frame.

---

281. $CH_3-\overset{O}{\overset{\|}{C}}-\overset{*}{C}H_2-\overset{*}{C}HO$ 281b: $Ph\overset{*}{C}H=CH-\overset{*}{\underset{\|}{C}}CH_3$ [cyclohexanone ring with *O and *CHO marked]

So at the moment, the scope of our skill at synthesis is rather limited. We can extend it by the use of α-β unsaturated ketones as electrophiles. Where would the compound we have just made (281b) be attacked by nucleophiles?

282. Have you shown two positions?

283.

$$PhCH=CH-\overset{O}{\underset{\underset{X^-}{}}{\overset{\|}{C}}}-CH_3 \qquad Ph\overset{\frown}{CH}=CH-\overset{O}{\underset{\underset{X^-}{}}{\overset{\|}{C}}}-CH_3$$

Fortunately, enolates usually prefer to add to the end of the double bond to give the more stable anion. Add the enolate from $CH_2(COOEt)_2$ to this ketone.

284.

$$EtO-\overset{O}{\overset{\|}{C}}-\overset{H}{\underset{}{CH}}=\overset{O}{\overset{\|}{C}}-OEt \quad \xrightarrow{^-OEt} \quad EtO-\overset{O}{\overset{\|}{C}}-CH=\overset{O^-}{\overset{|}{C}}-OEt$$

$$PhCH=CH-\overset{O}{\underset{}{\overset{\|}{C}}}-CH_3$$

$$\downarrow$$

$$(EtO_2C)_2CH-\underset{\underset{Ph}{|}}{CH}-CH=\overset{O^-}{\underset{}{\overset{|}{C}}}-CH_3$$

This intermediate is of course another, less stable, enolate anion, and in the ethanolic solution it will protonate to give the final product. Draw this.

285.

$$(EtO_2C)_2CH-\underset{\underset{Ph}{|}}{CH}-CH=\overset{O^-}{\overset{|}{C}}-CH_3 \quad \xrightarrow{H-OEt} \quad (EtO_2\overset{*}{C})_2CH-\underset{\underset{Ph}{|}}{CH}-CH_2-\overset{O}{\underset{*}{\overset{\|}{C}}}-CH_3$$

Notice that the compound we have just made has a 1,5(*) relationship between the two carbonyl groups. This is an example of the Michael reaction. Here is another: suggest what the product would be.

[cyclopentanone with CO₂Et substituent at α-position] + $CH_3CH=CH-CHO$ $\longrightarrow$

286. 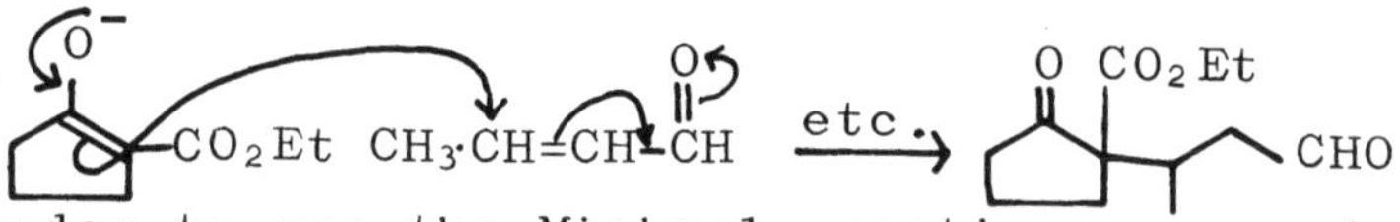 In order to use the Michael reaction, we need to be able to make α,β-unsaturated carbonyl compounds of various types, but we already know how to do this. How would you make this one:

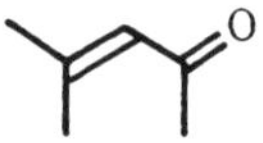

287. By the self condensation of acetone:

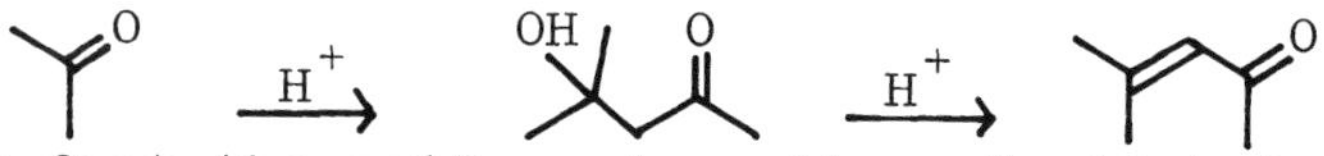

In fact the self condensation of aldehydes and ketones is a good way to make Michael reagents. We can't however make α-β unsaturated esters this way. How can we make this, then:

$$PhCH=CHCO_2Et$$

288. There are three methods which we have met so far:

i. Ester and carbonyl compound which can't enolise:

$$PhCHO + CH_3CO_2Et \xrightarrow{EtO^-} PhCH=CHCO_2Et$$

ii. Perkin reaction: (frames 252-7)

$$PhCHO + (CH_3CO)_2O \xrightarrow{AcO^-} PhCH=CH.CO_2H$$

$$\xrightarrow[H^+]{EtOH} PhCH=CHCO_2Et$$

iii. Organo-zinc reagents (frames 208-212):

$$CH_3CO_2H \xrightarrow{P, Br_2} BrCH_2COBr \xrightarrow{EtOH} BrCH_2CO_2Et$$

$$\xrightarrow{Zn} BrZnCH_2CO_2Et$$

$$PhCHO + BrZnCH_2CO_2Et \longrightarrow PhCH=CHCO_2Et$$

289. A compound we have mentioned but not used is nitromethane (frame 262), a good source of a stable enolate. What product would we get here:

$$CH_3NO_2 + PhCH=CHCO_2Et \xrightarrow{EtO^-} ?$$

290.

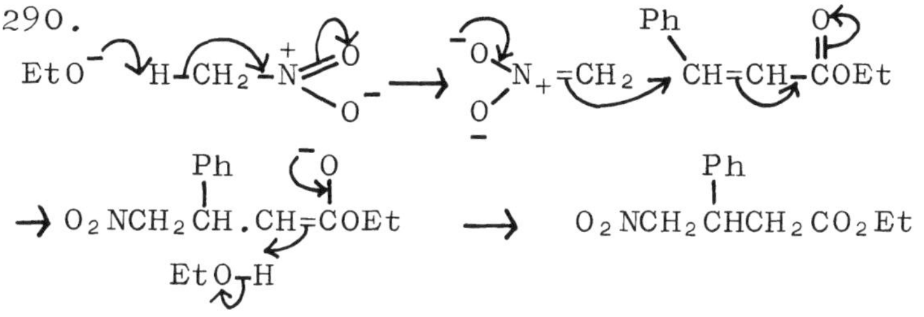

EtO⁻ ⤴ H–CH₂–N⁺(=O)(O⁻) → ⁻O–N⁺(=CH₂)(O⁻) , Ph–CH=CH–COEt

→ O₂NCH₂CH(Ph).CH⁻–COEt , EtO–H → O₂NCH₂CH(Ph)CH₂CO₂Et

---

291. Another non-carbonyl activating group is cyanide, which withdraws electrons in much the same way as carbonyl. Acrylonitrile, $CH_2=CHCN$ is a useful Michael reagent. What would happen here:

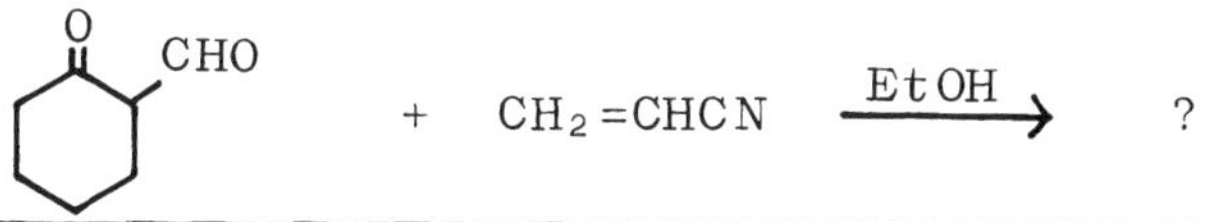

cyclohexanone-2-CHO + $CH_2=CHCN$ $\xrightarrow{EtOH}$?

---

292. Michael addition of enolate anion to acrylonitrile:

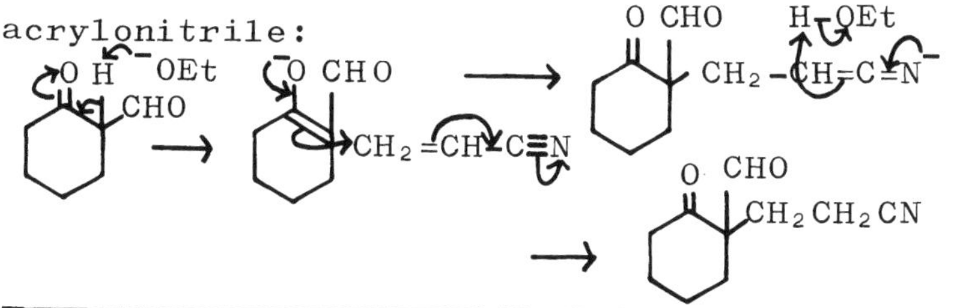

[mechanism diagram showing enolate formation, addition to CH₂=CH–C≡N, and product 2-(CHO)-2-(CH₂CH₂CN)cyclohexanone]

---

293. To finish off the Michael reaction, let's take two compounds we've already made (frames 230-232) and combine them together:

294. Yes, I know this is a very complicated example, but use your knowledge: remove a proton from the most acidic position, and add the resulting anion Michael fashion to the other compound.

This example is fictional, that is to the best of my belief no one has tried it, and maybe you don't blame them. I put it in just to show you the complicated molecules which can be made this way. Just suppose someone asked you how to make this molecule: I imagine you would find it a rather daunting problem! Yet these are reactions and compounds we have ourselves made in this program with straight-forward ideas.

This is a good place to rest if you want to.

295. It must be clear to you by now that the carbonyl group not only allows certainly reactions to occur which would otherwise be impossible: it also directs where they will occur by selecting exactly which hydrogen atom shall be removed in enolisation and therefore exactly where the new carbon-carbon bond is to be

formed.

Organic chemists often introduce "unnecessary" carbonyl groups into molecules during a synthesis, just so that they can use this direction finding function. These groups are not wanted in the final molecule and so we must be able to remove them later.

We have just been discussing this molecule:

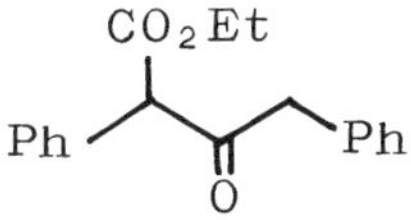

Ph-CH(CO₂Et)-CO-CH₂-Ph

What would happen if we boiled this up in water containing a catalytic amount of acid?

---

296. The ester group would hydrolyse:

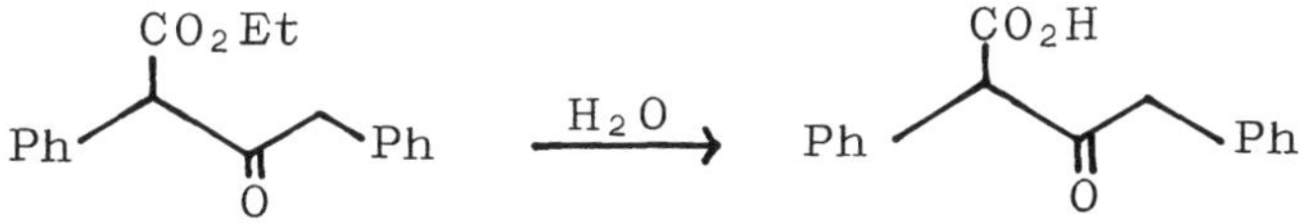

Ph-CH(CO₂Et)-CO-CH₂-Ph →(H₂O)→ Ph-CH(CO₂H)-CO-CH₂-Ph

and as soon as the free carboxyl group was released, it would be released in another sense in that it would form CO_2. Arrows?

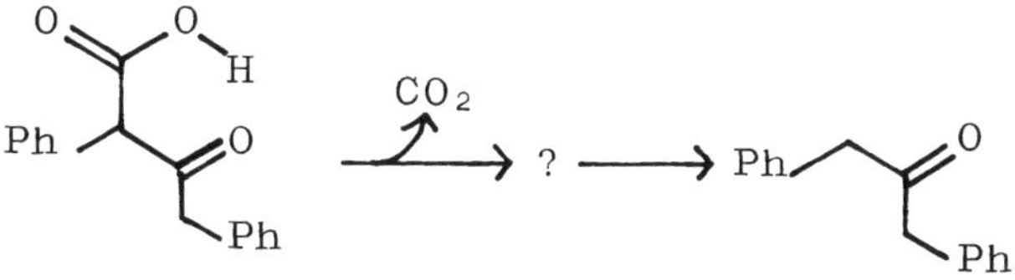

[structure with cyclic H-bonded carboxyl] → (−CO₂) → ? → Ph-CH₂-CO-CH₂-Ph

---

297. The missing intermediate is the enol:

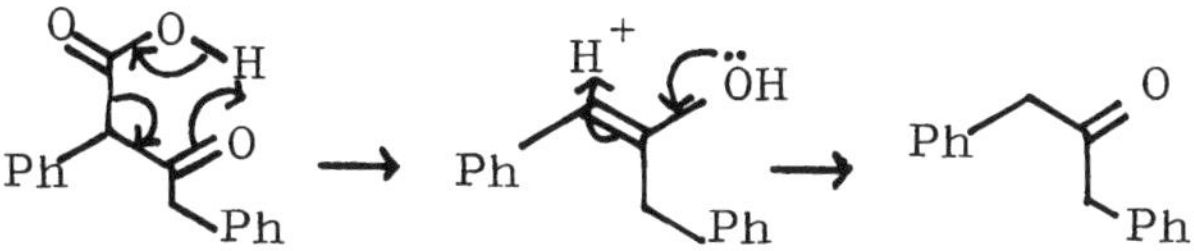

[enol intermediate structures] → Ph-CH₂-CO-CH₂-Ph

The decarboxylation reaction happens to this sort of compound too:

From frame 285:

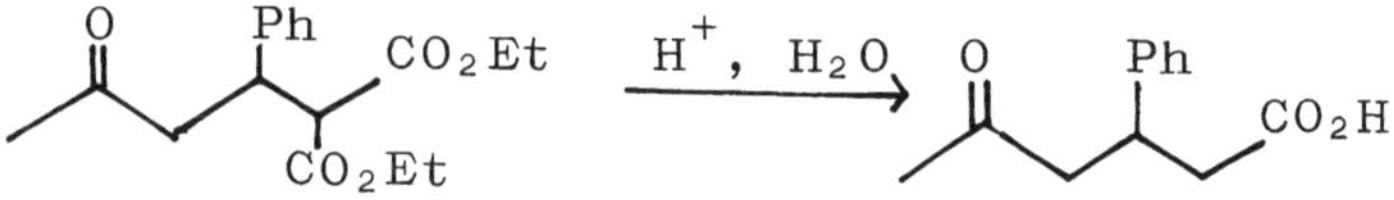

CH$_3$-CO-CH$_2$-CH(Ph)-CH(CO$_2$Et)$_2$ →[H$^+$, H$_2$O] CH$_3$-CO-CH$_2$-CH(Ph)-CH$_2$-CO$_2$H

Mechanism?

298.

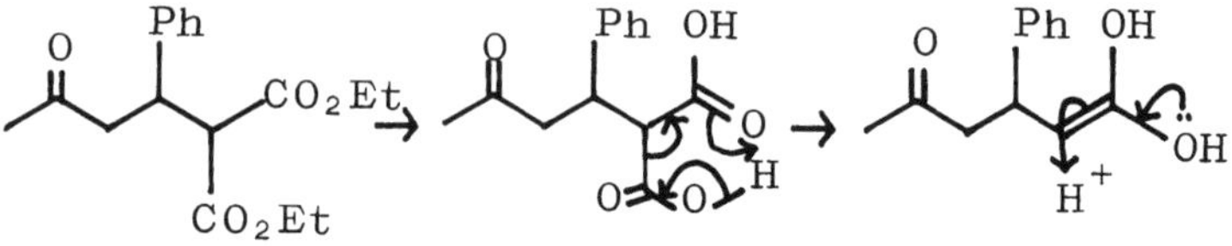

Hydrolysis of both ester groups is followed by decarboxylation of one of them to give the enol of the final product.

Decarboxylation is a useful addition to the alkylation reactions we mentioned in frames 275–8. What happens here:

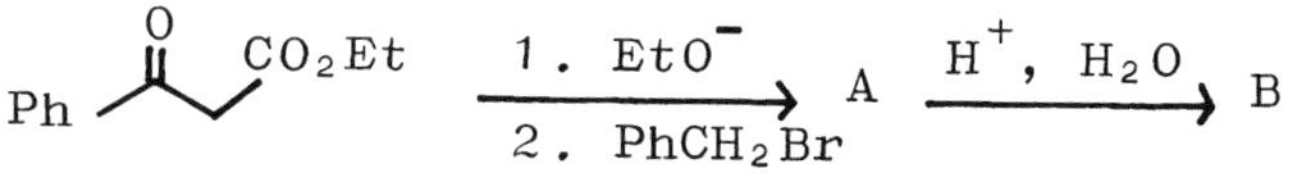

Ph-CO-CH$_2$-CO$_2$Et →[1. EtO$^-$; 2. PhCH$_2$Br] A →[H$^+$, H$_2$O] B

---

299. First we alkylate the enolate anion:

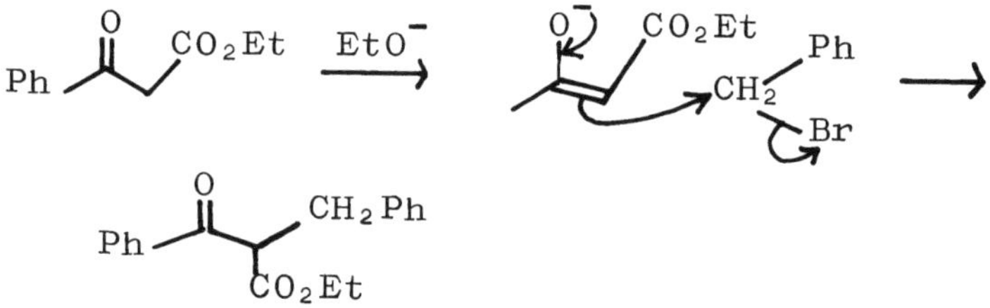

Ph-CO-CH$_2$-CO$_2$Et →[EtO$^-$] enolate + PhCH$_2$Br → Ph-CO-CH(CH$_2$Ph)-CO$_2$Et

then hydrolyse and decarboxylate the ester:

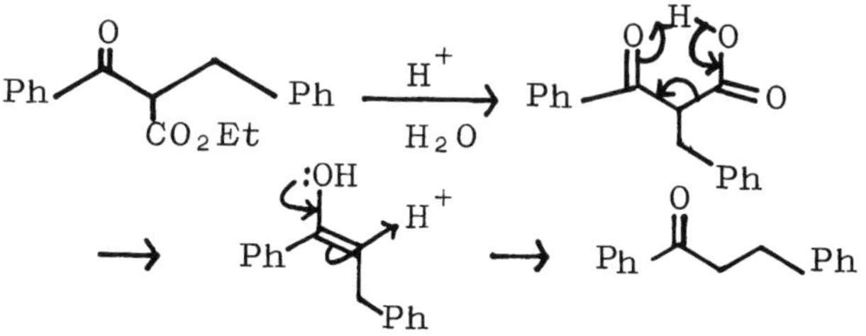

Ph-CO-CH(CO$_2$Et)-CH$_2$-Ph →[H$^+$, H$_2$O] cyclic transition state → Ph-C(OH)=CH-CH$_2$-Ph →[H$^+$] Ph-CO-CH$_2$-CH$_2$-Ph

300. If we combine decarboxylation with the Michael reaction (frames 281-5) we get a general synthesis of 1,5-diketones:

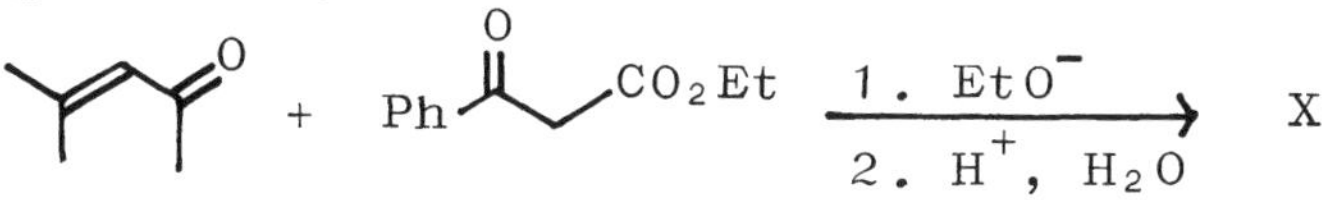

CH₃-C(=CH)-C(=O)-CH₃ + Ph-C(=O)-CH₂-CO₂Et →[1. EtO⁻][2. H⁺, H₂O] X

301. The β-dicarbonyl compound will enolise, as usual, and add to the Michael reagent:

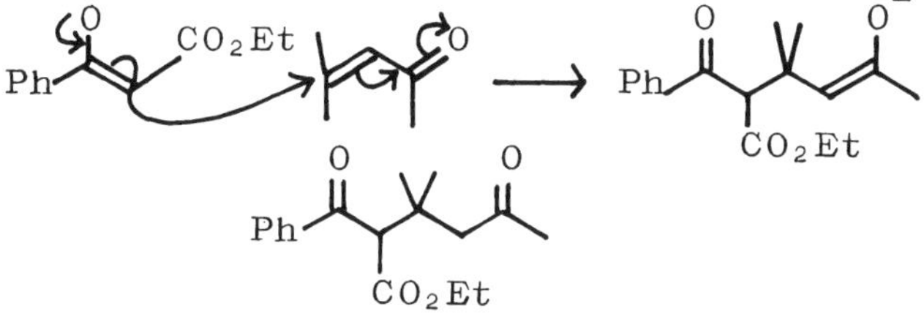

Ph-C(=O)-CH(CO₂Et) ... → Ph-C(=O)-CH(CO₂Et)-CH₂-C(CH₃)₂-C(=O)⁻ ...

Ph-C(=O)-CH(CO₂Et)-CH₂-C(CH₃)₂-C(=O)-CH₃

Now we must hydrolyse the ester and the usual decarboxylation follows:

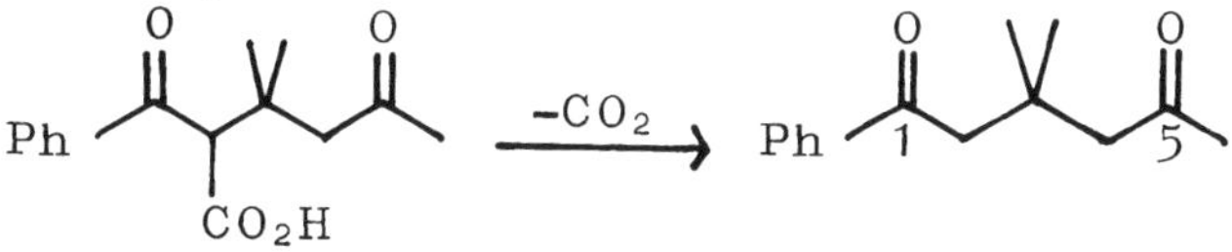

Ph-C(=O)-CH(CO₂H)-CH₂-C(CH₃)₂-C(=O)-CH₃ →[-CO₂] Ph-C(=O)-CH₂-CH₂-C(CH₃)₂-C(=O)-CH₃
                                                     1   5

302. A more deep-seated change than decarboxylation can easily occur. You will remember that I have stressed all along that these carbonyl reactions are reversible, though we don't always show them as such. Even the formation of carbon-carbon bonds is reversible. If we take some of this compound and treat it with EtO⁻, will it form a substantial amount of enolate?

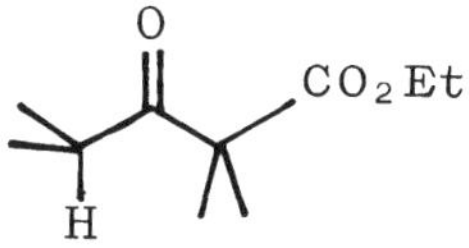

(CH₃)₂CH-C(=O)-C(CH₃)(H)-CO₂Et

303. No, because although it can form an ordinary enol: the two carbonyl groups can't cooperate in forming a very stable enol as the position between them is blocked. What happens instead is that EtO⁻ attacks the more reactive carbonyl group to give an adduct. Draw this.

304. The ketone carbonyl is more reactive than the ester carbonyl:

The adduct then decomposes with the loss of the enolate anion of the ester as a leaving group. Draw this.

305.

If the compound had been able to form a substantial amount of enol, it would have done this much less readily. We can generalise from this reaction and say that under the right conditions we ought to be able to "undo" any of the C-C bond forming reactions we have been using. What would happen here:

306. This time the aldehyde is the more reactive group:

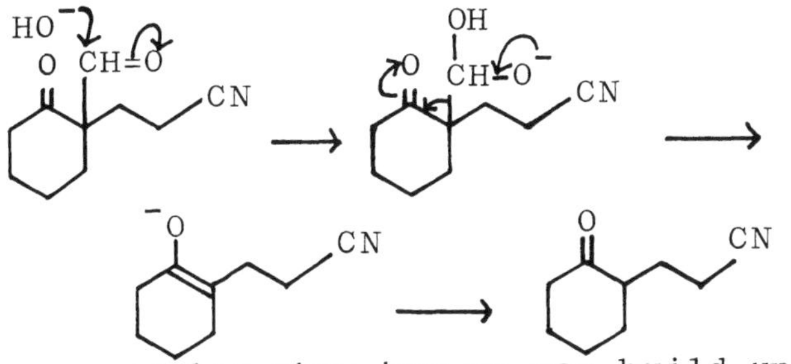

These examples show how we can build up a molecule in disguise and reveal it at the last minute by a specific cleavage reaction. The point of leaving the cleavage reaction until the end is that we retain the 1,3-relationship of the two carbonyl groups and can thus control the way the molecule reacts. These cleavage reactions are even more dramatic when they are applied to cyclic compounds, and it is time for us to look at the formation and cleavage of rings in more detail.

---

307. We have met six membered rings several times during the program, and noted their stability. This is a property they share with five membered rings.

We have also met this β-keto ester, but we haven't yet worked out how we would synthesise it. Suggestions?

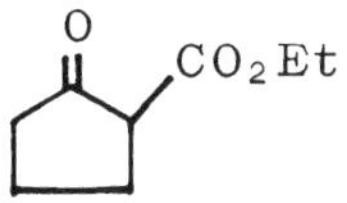

308. We made β-keto esters in the past by the self condensation of esters (frames 221-230). This molecule is also made by the self condensation of an ester.

---

309. The reaction is intra-molecular: the enolate component being one end and the electrophilic component the other end of a six carbon diester:

[structure: EtO-C(=O)-(CH2)4-C(=O)-OEt →(EtO⁻) cyclized intermediate → cyclopentanone with CO2Et substituent]

---

310. In cyclic compounds, the problem of ambiguous reactions often just doesn't arise because the molecule will always prefer reactions which form five or six membered rings to those which form other sizes. Four membered rings are particularly bad. Take this diketone, for example:

[structure: cyclohexanone with -CH2CH2C(=O)CH3 side chain] →(EtO⁻) ?

it can form four different enolates, each of which could react with the other carbonyl group, and yet only one product is formed. Which?

---

311. Two possible reactions give four membered rings, and can be discounted, one reaction gives a good stable six membered ring. There is actually another possibility which you could discuss with your adviser if you are

interested. It is beyond our scope now.

312. Quite often these cyclisation reactions happen spontaneously after some other normal reaction. Carry out the normal Michael reaction between these two compounds and then see if you can predict what cyclisation will follow:

$$\text{(structure)} + CH_2(CO_2Et)_2 \xrightarrow{EtO^-} ?$$

313. Here is the Michael reaction:

$(EtO_2C)_2\bar{C}H$ ... → $(EtO_2C)_2CH$ ...

→ $(EtO_2C)_2CH$ ...

and here is the cyclisation: the only one to give a six membered ring. How could we remove the COOEt group from this product, and what would we get?

314. By acid hydrolysis: decarboxylation would follow:

[EtO₂C-substituted dimedone] →(H⁺/H₂O) [HO₂C-substituted dimedone] → [dimedone]

Try your hand at another reaction like this: a reaction between two compounds which is followed by cyclisation; this time a five membered ring is formed:

Ph-CO-CO-Ph + Ph-CH₂-CO-CH₂-Ph →(EtO⁻) ?

---

315. Only one compound can enolise, and the diketone will act as the electrophilic component as it is more reactive than a mono ketone. We begin with a simple condensation:

[mechanism: deprotonation by ⁻OEt, enolate attacks diketone, giving PhCH(Ph)-C(OH)(Ph)-CO-Ph intermediate]

Now we are all set up for an intramolecular condensation:

[cyclisation to diol, then −2H₂O gives tetraphenylcyclopentadienone]

---

316. The cleavage reactions we talked about a few frames ago (302-6) can also occur in cyclic compounds and we may get ring opening here? What might this lead to:

[2-benzoyl-1-(ethoxycarbonyl)cyclopentane with O Ph, CO₂Et] →(EtO⁻) ?

317. The ketone carbonyl is the more reactive so EtO⁻ adds there. The only leaving group is the ester enolate:

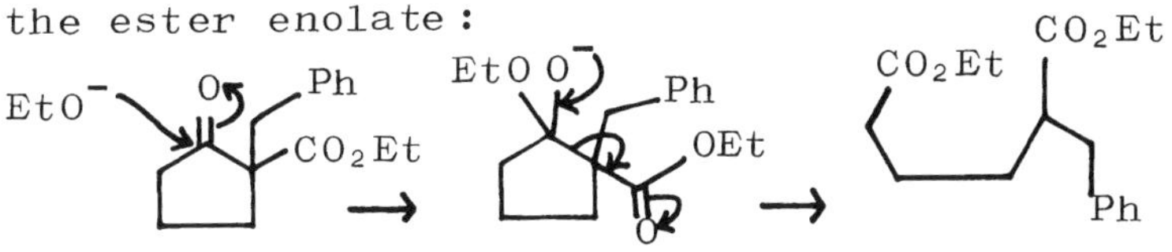

318. Sometimes ring opening can be caused by other reactions than base cleavage. Predict what happens here in a reaction which contains a bit of revision work:

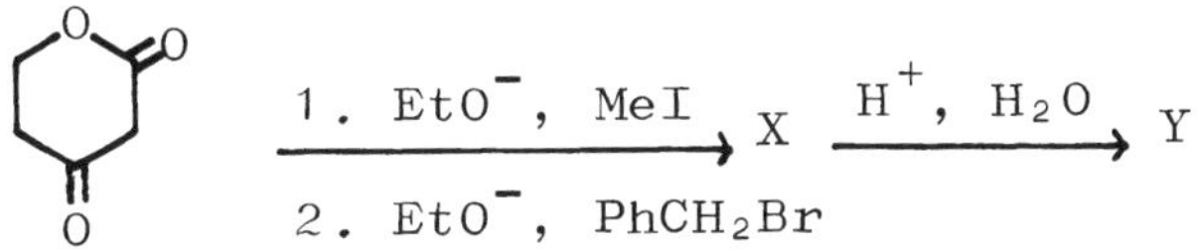

$$\text{(lactone-ketone)} \xrightarrow[\text{2. EtO}^-, \text{PhCH}_2\text{Br}]{\text{1. EtO}^-, \text{MeI}} X \xrightarrow{\text{H}^+, \text{H}_2\text{O}} Y$$

319. The first two steps are simply alkylations (frames 275-8):

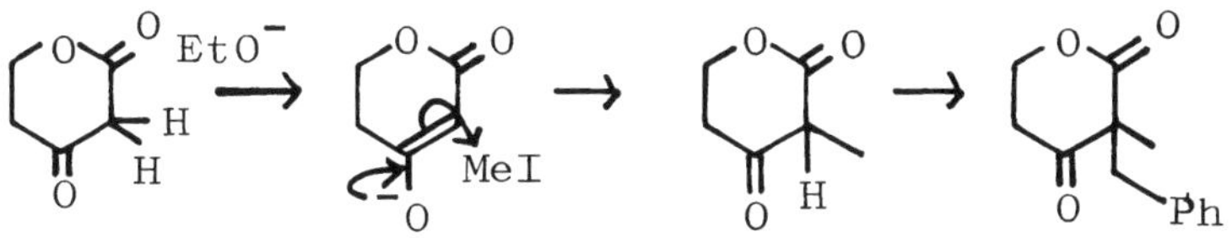

Then acid hydrolysis opens up the cyclic ester, and decarboxylation follows:

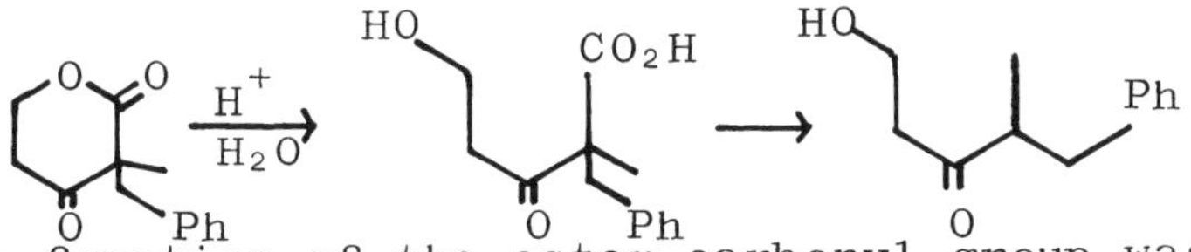

The function of the ester carbonyl group was simply to introduce the two alkyl groups: after it had done this it could easily be removed.

This is a good place to stop and rest if you want to.

320. We have now completed the exploration of new territory and we should stop here a moment to consolidate the new ideas. The main ideas have been ways to make new carbon-carbon bonds between two compounds, both having a carbonyl group, but each acting in a different way.
What do we call these two reagents?

---

321. One is the nucleophilic or enolic component. The other is the electrophilic or carbonyl component. We have discovered various ways to join these components unambiguously. Make a list of these.

---

322. Have you remembered to include reactions in which both components are the same molecule and reactions which form rings?

---

323. One way to express this is:
a) Self condensation: both components derived from the same molecule (frame 230).
b) Electrophilic component incapable of enolising and more reactive than the enolic component towards nucleophilic attack. (frames 240-244).
c) Enolisable component is very reactive in the sense that substantial amounts of enol are formed (mostly β-dicarbonyl compounds) but unreactive in the electrophilic sense. (frames 260ff).
d) A six or five membered ring is formed. Other possible reactions lead to less stable

rings of other sizes (often four membered). (frames 310-311).

You may well have expressed this just as correctly but in a different way. If you are in doubt, discuss with your adviser.

---

324. The rest of the program will try to put these ideas into perspective by looking at some actual syntheses, offering some general problems, and generally giving you a hint of the enormous possibilities in these simple ideas. Don't be put off if you find the going a bit rough: you've really done all the hard work by this stage!

Here is a step from the synthesis of a steroid - one of a large family of physiologically active molecules including hormones. Explain what is happening.

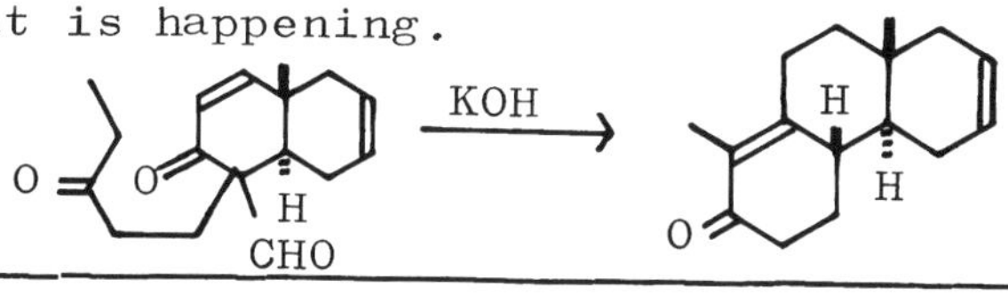

---

325. It is clear that the CHO group has been lost, presumably in a β-dicarbonyl cleavage reaction (305-6), and that a condensation has taken place to form a new six membered ring. The order of these two reactions doesn't matter much;

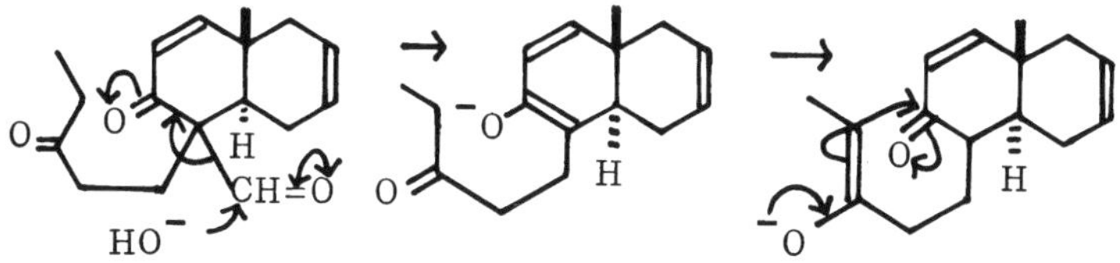

326. Here is part of another steroid synthesis. This time, see if you can suggest how to carry out the steps marked *.

327.

BUILDING ORGANIC MOLECULES, 5:38

A → B: cross condensation of type 323b.
B → C: simple alkylation of stable enolate anion.
D → E: hydrogenation of a double bond.
E → F: steps outside the scope of this program
F → G: cyclisation to form stable five membered ring.
G → H: hydrolysis and decarboxylation.

You may be interested to see the full steroid skeleton, and here it is. Perhaps you can see that these two syntheses are moving towards it, though both still need many steps.

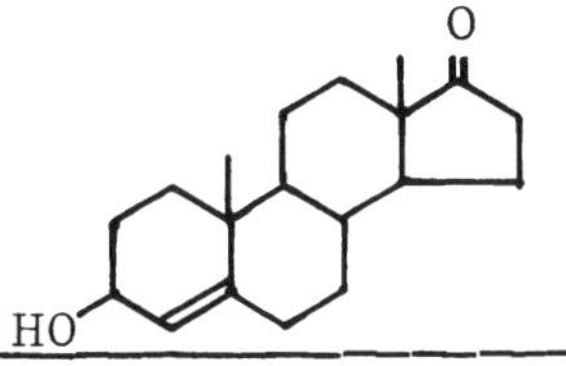

a steroid

---

328. I shall end the program with one complete synthesis: that of cedrene, the essential oil of cedar wood. The synthesis has many steps and requires some reactions you haven't met, but I thought you would like to see the carbonyl group in action. Here it is:

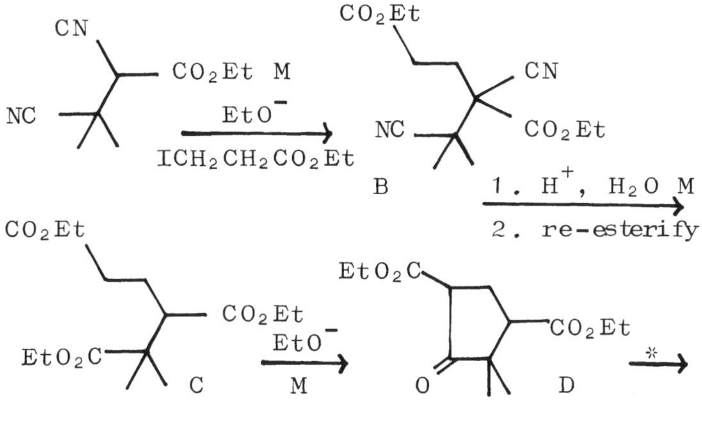

BUILDING ORGANIC MOLECULES, 5:39

You should be able to suggest reagents for the steps marked * and mechanisms for the steps marked M. Don't worry about the other stages in the synthesis, they use reagents you haven't met.

---

329. Here are outline answers: if you're in doubt about any of them, discuss them with your adviser.

A → B: Alkylation of a stable enolate.
B → C: Decarboxylation after ester hydrolysis.
C → D: Cyclisation to form a five membered

ring: all the other possible reactions
give three or four membered rings.

D →E: Make the stable enolate with EtO⁻ and
alkylate with MeCHBr.COOCH₂Ph.

F →G: The methyl group on the ketone carbonyl
is the enolate component attacking the
ring ketone and forming a stable five
membered ring.

H →I: Make the cyclic ketal with HSCH₂CH₂SH
and hydrogenate over Raney nickel
(frame 146).

J →K: A fine reaction, this! The methyl group
on the ketone makes an enolate, and this
attacks the ester group in a substitution reaction. Though the new ring looks
a bit awkward, it is in fact a stable
six membered ring.

330. We stopped this synthesis when we had
made the cedrene skeleton, and there are several stages still left before cedrene itself
is formed. Cedrene is in fact:

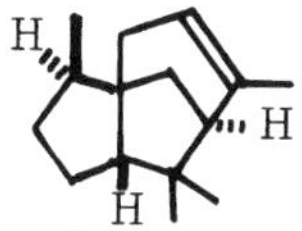

so that all the carbonyl groups used during
the synthesis, a small matter of four ester
functions and three ketone functions, were put
in to guide the synthesis along - the final
product contains no carbonyl groups at all!
This synthesis was designed by Professor
Gilbert Stork and the genius in the plan is
clear: every time he wants to make a carbon-
carbon bond he has carefully arranged for the

presence of two carbonyl groups, one to stabilise the enol and one to be the electrophile. These are the principles we have been studying here.

This is the end of the program.

INDEX

The numbers are FRAME numbers.

The index includes names of individual compounds, classes of compounds, reaction types, and named reactions.

Aldehyde, ketone, etc. are not indexed as these compounds appear on almost every page.

Concepts and ideas are better found from the pages at the front of each section.

INDEX

Acetaldehyde: 5, 17, 29, 38, 134-5, 174, 220, 258-9, 280
Acetals: as protecting groups: 64-5
    cyclic: 62-4
    synthesis: 3-18, 62-5
    hydrolysis: 18-21
Acetamide: 75-6, 94-7, 115, 215-6
Acetate ion as catalyst: 253
Acetic acid: 288
Acetic anhydride: 75-6, 115, 215-6, 254, 288
Acetone: 3, 5, 7, 29, 51, 55, 64, 121-5, 130-2, 169-173, 208-9, 228, 270-3, 280
Acetophenone: 29, 45, 127-9, 189-201, 211, 241, 250, 262
Acetyl chloride: 75-6, 115, 215-6
Acetylene: 31
Acetylenes, synthesis: 153-4, 167-8
Acetylide anion: 31-2
Acid catalysis, in acetal synthesis: 5-15, 62-4
    hydrolysis: 18-9, 64-5
    amide hydrolysis: 94-101
    aromatic electrophilic substitution: 156-162
    bromination of ketones: 198-201
    condensations: 227-9, 232, 240-1
    dehydration of alcohols: 40-1, 83, 164, 210
    ester hydrolysis: 314, 319
    choice of acid in: 162-3

INDEX

Acid anhydrides: synthesis: 75-6, 113-4
    reactions: 252-6
Acid bromides: 202-212
Acid chlorides: and acid anions: 75-6, 114
    and amines: 71-2
    esters from: 111
    hydrolysis: 66-70
    synthesis: 102-6
Alcohols: dehydration: 39-40, 82-6
    oxidation: 60-1
    reactions: 5, 9-21, 39-41, 51, 83-6, 111-3, 120, 229
    synthesis: 23, 31-2, 35-9, 44-5, 76-83, 208-9, 220, 228, 280
Aldol reaction: 219-220
Alkylation of enolate anions: 275-8, 299, 319, 329
    by Michael reaction: 284-294
Aluminium: 45-6
    hydride ion: 41-4, 76-9, 120, 139
    isopropoxide: 47-53
    tert-butoxide: 60-1
Amide ion: 31
Amides: hydrolysis: 94-102
    synthesis: 71-2, 75-6
Amines: and esters: 71-2, 75-6
    and ketones: 123-5
Aniline: 71-2
Azines: 131-2

INDEX

Base catalysis: choice of 224-6, 262
    in amide hydrolysis: 94
    in bromination of ketones: 189-197
    in condensation reactions: 219-226
    in elimination reactions: 152-4, 168
    in enolisation: 173
    in ester hydrolysis: 89-93
    in esterification with acid chlorides: 111
Benzaldehyde: 19, 29, 165-6, 237, 253-6, 274, 280, 288
Benzene: 160
Benzil: 315
Benzophenone: 29, 39, 258-9
Benzoyl chloride: 66-72
Benzyl alcohol: 160-1
Benzyl bromide: 278, 318-9
Benzyl chloride: 161
Borohydride ion: 44-5, 55, 139
Boron: 45-6
Bromination
    of acids: 202-204
    of ketones: 189-201
Bromine: 189-207
Bromoform reaction: 189-197
n-Butanol: 17

Carbanions, simple: 169-170, 195-6
    see also enolate anions
Carbonium ions: 40, 137-8, 83, 149-150, 157-162, 164, 210

INDEX

Cedrene, synthesis: 328-330
Chloral: 156-8
Chloride ion as nucleophile: 149-152, 162-3
Chlorobenzene: 156-8
Chloroform reaction: 197
Chloroformate ester: 327
Chloromethylation: 159-161
Claisen ester condensation: 221-3, 277
Cleavage of 1,3-dicarbonyls: 304-6, 325
Clemmensen reduction: 140
Components in condensations: 220, 230, 235, 260, 271, 309, 315, 320-323
Condensations:
    cross-: 235, 240-6, 260
    self-: 230-232, 236-240, 277, 287, 308-9, 323
Conjugation: 26, 176, 183
Crotonaldehyde: 285-6
Cyanide
    electron-withdrawing substituent: 269, 291
    ion, as nucleophile: 7-8, 23-5, 83
Cyanoacetic ester: 268
Cyanohydrins, synthesis: 22-30
Cyclic acetals: 62-4
Cyclic compounds, effect of ring size: 167-8, 310-1, 323
Cyclohexanone: 60, 142-3, 232, 244-5, 280
Cyclononyne: 167-8

INDEX

DDT: 158
Decarboxylation: 296-301, 313-4, 327, 329
Dehydration of alcohols: 39-40, 82-6, 209-212
1,3-(β-)Dicarbonyl compounds: 260, 267
    alkylation: 275
    cleavage: 302-6, 325
    Michael addition with: 284
    synthesis: 280-1
1,5-Dicarbonyl compounds: 285, 300
gem-Dichlorides from ketones: 149-152, 167-8
Dieckmann reaction: 307-9
1,3-Diketones: 174, 213, 262
Dimedone synthesis: 314
2,4-Dinitrophenylhydrazine: 165-6
Dithioacetals: 144-6, 148, 329

E2 mechanisms: 154
Elaboration in synthesis: 233
Electrophilic aromatic substitution: 155-162
Electrophilic component: 220
Elimination reactions: 39-48, 82-6, 153-4, 164, 167-8, 256
Enol,
    compounds which cannot form an: 237-240
    conjugated: 175-6
    intermediate in decarboxylation: 297
    reaction with bromine: 198-201
    synthesis: 172-3
Enolate anion: 177
    alkylation: 275-8, 299, 329
    as leaving group: 304

INDEX

      cis-trans equilibration: 188
      conjugated: 179
      delocalisation in: 171
      pK_a: 262-7
      planarity: 186
      reaction with bromine: 198-201
      synthesis: 170
Equilibrium in addition to carbonyls: 25-30
Esters,
      alkaline hydrolysis: 89-93
      mechanism of hydrolysis with $H_2{}^{18}O$: 117-8
      reactions: 71-5, 76-9, 80-8
      synthesis: 110-3, 120
Ethyl acetoacetate: 213
Ethylene glycol,
      acetals from: 62-5
      solvent: 143

Five-membered rings,
      cleavage: 317
      reactions: 275, 285
      synthesis: 62, 307-9, 315, 327-9
Formate esters: 237, 244-5, 280
Formaldehyde: 35-7, 160-1, 237, 248-251, 262
Four-membered rings, not formed: 255, 310

Glycol, ethylene: 62-4, 143
Grignard reagents,
      and esters: 82-8
      and ketones: 35-40
      mechanism of reaction: 87-8

INDEX

Hell-Volhard-Zelinsky reaction: 204, 212
Hydrazine: 130-133, 141-2
Hydrazones,
      reaction with base: 141-3
      synthesis: 130-1
Hydride donors: 41-5
Hydride transfers: 52-3
Hydrochloric acid: 160-3
Hydrogen bonding: 176, 214
Hydrogenation: 146, 327, 329
Hydrolysis,
      of acetals: 18-21
      of amides: 94-102
      of esters: 89-93, 295-301
Hydroxylamine: 126-7
Hydroxymethylene compounds, synthesis: 245

Inductive effects: 26, 194
Imines, synthesis: 121-7
Intramolecular reactions: 53, 64, 87-8, 296-9,
      308-315, 323-9
Iodoform reaction: 197

β-Keto acids, decarboxylation: 296-301
β-Keto esters,
      alkylation: 275-8
      enolisation: 213, 262, 267, 275
      synthesis: 221-3, 230-1
Keto form: 172
Knoevenagel reaction: 273-4

INDEX

Lactones: 85-6, 318-9
Leaving group: 68-9, 95-7, 121-5
    and basicity: 72-5, 78
    enolate ion as: 304
Lithium aluminium hydride: 41-4, 76-9, 120, 139-140, 234
Lone pair electrons: 27

Malonate esters: 261-2, 284, 312
    decarboxylation: 298
Mannich reaction: 248-251
Meerwein-Pondorff reduction: 47-53, 65
Mesityl oxide: 287-8, 312
Methyl iodide: 275, 300
Michael addition: 281-294, 300-1, 312-3
Monobromination of ketones: 198

Nine-membered ring: 167-8
Nitriles: 262, 291, 328
Nitro compounds: 240-1, 257, 262, 266, 289
p-Nitro-benzaldehyde: 240-1, 256-7
Nitromethane: 289-291
Nucleophilic component: 220

Olefins, synthesis: 39-40, 82-86, 164
Oppenauer oxidation: 60-1
Oximes: 126-9
Oxygen, labelled (^{18}O): 117-8

Perkin reaction: 252-7, 288
Phosphorus pentabromide: 202
Phosphorus pentachloride: 150-1, 167-8

INDEX

Phosphorus, red: 204-212
pK_a: 33-4, 262
Potassium bisulphate: 83-6
iso-Propanol: 51

Racemisation: 186
Raney nickel: 146
Reactivity of carbonyl compounds: 25-30, 79, 115-6, 215-6, 304, 306, 317
Reduction
    aldehydes: 78-79
    esters: 76-9, 234
    ketones: 42-5, 47-59, 136-148, 234
Reformatsky reaction: 208-212, 288, 327

$S_N 2$ reaction: 276
Secondary amines as catalysts: 248-251, 270-1
Self-condensations: 230
    of aldehydes: 219-220
    of esters: 221-6, 230-1, 276-8, 309
    of ketones: 228-9, 231-2, 287
Semicarbazide: 133-5
Six-membered rings,
    cleavage: 84-6
    reactions: 60, 84-6, 142-3, 206-7, 213-4, 232, 244-5, 250-1, 280, 291-4, 305-6, 310
    synthesis: 310-4, 325, 329
Sodium bicarbonate: 114
Sodium borohydride: 44, 55
Steroid syntheses: 324-7

INDEX

Stork, Professor Gilbert: 330
Sulphuric acid: 162-3

Tautomerism: 172
Tetracyclone, synthesis: 315
Thiols: 7-8
Thionyl chloride: 103-6, 120
Toluene-p-sulphonic acid: 17
Transition states,
    how to draw: 54-8
    six-membered cyclic: 59, 88, 255

α,β-Unsaturated acid derivatives, synthesis:
        209-212, 252-7
α,β-Unsaturated aldehydes: 285-6
α,β-Unsaturated esters, synthesis: 273-4,
    287-8, 327
α,β-Unsaturated ketones:
    as electrophiles: 281-294, 312-4
    enols from: 177-9
    synthesis: 227-9, 231-2, 241, 280, 286-8, 315, 324-5, 328
β,γ-Unsaturated ketones: 179-180, 184
α,β-Unsaturated nitriles: 291-2

Wolff-Kishner reduction: 141-3, 148

Zinc amalgam: 140
Zinc, organo-, reagents: 208-212, 288, 327